THE CRYPTOCURRENCY SHIFTERS

Learn How To Start Trading In Cryptocurrency World

EMIL COLLINS

Table of Contents

Introduction

Cryptocurrency and bitcoins are possibly two of the most buzzing words in today's technological and business landscape. And for a good reason too! If recent statistics are to be believed, then the popularity of Cryptocurrencies and Bitcoins is at an all-time high!

Believe it or not, a post about Bitcoin is made on social media every 3 seconds! Not only that, the price of Bitcoin has surged to over 66%, and the market capitalization holds a value of 1072.21 billion and still growing! It is expected that the market size of the global blockchain industry will grow up to 23.3 billion by the end of 2023.

Those numbers are astounding because Cryptocurrency is still a relatively new concept from a technological and a financial standpoint.

If you've visited a financial website in the last few years, you've probably heard of Bitcoin. Cryptocurrencies, block-chain, Ethereum, portability, fungibility, and other esoteric terms were once only known to those involved with marketing or finances. know. All of them are now being commonly used, and it is easy to become intimidated by this brave new world of investing.

Most people are interested in cryptocurrencies these days because they've heard about the unfathomable sum of money that can be made in the market. What was once a niche asset reserved for computer scientists has evolved into a fully viable alternative asset that has been compared to gold, oil, and several other mainstream assets.

This book aims to demystify this asset class, particularly Bitcoin, and help you determine whether this investment is a good fit for you. Even if you do not intend to invest, it is worthwhile to understand what forces drive this new asset class and how long these forces will last. Many people do not understand the economic forces currently driving the prices of Bitcoin and other cryptocurrencies.

On the other hand, cryptocurrencies are here to stay, and ignoring them will not make them go away or help you understand them. I've had a front-row seat to the evolution of money as someone who has worked in financial services technology for over 30 years. All of the viewpoints expressed in this book are the same as those I advocate for my clients worldwide.

So, you can already tell that I have a very good grasp of the overall concept of Cryptocurrency and understand just “Why” it is going to be the next big thing in the world of finances, marketing, and technology.

If you've decided to buy this book, I'm assuming you're new to this world. While this book will not turn you into a trading genius, it will help you get past the beginner stage.

Cryptocurrency is a disruptive concept which is used as an alternative to fiat currency in the current monetary system. Entrepreneurs, startups, and large, small, and medium-sized businesses are interested in cryptocurrencies. They see it as a game-changing tool for combating transactional compliances.

As the cryptocurrency market expands, there is an increasing global demand for blockchain technology developers. Companies are looking for people who can design software, contracts, applications, and other related items in this field. If you understand programming fundamentals, cryptocurrency could be the start of a lucrative career for you.

If you follow the values outlined in this book, it can guide your success. Because of the world of cryptocurrency, many people are now living unimaginably good lives. You have a good chance of being one of them.

The cryptocurrency market is expected to be extremely profitable for those who understand how it operates. You can achieve success and financial freedom with the right knowledge, as provided and revealed in this book, with a consistent hard work.

It will never be easy to make money in the financial markets. It will be even more difficult for newcomers. It can be harsh and perplexing. One thing is certain: it is not a get-rich-quick scheme. In this market, such a mindset will do nothing but drown us.

You may be familiar with a few financial market terms, but you may be new to trading. If you enter the financial markets with the mindset of a gambler, expecting to make a lot of money in a few days, then it is a bad habit which you should break if you want to be a successful trader.

If the first thing you do after winning trade is to increase your trade amount the next time, or if you double down when you lose money to make it all back, you have a gambler's mentality which you should break. This book ensures that you are not burned too early in your trading career. It aims to ensure that you will no longer think like a newbie by the end of this book.

Many people invest in the development of payment gateways and websites for their currencies. When a customer makes a cryptocurrency purchase, the transaction frequently passes through a fixed exchange rate through the payment gateway. It instantly converts to a widely accepted fiat currency, allowing the merchant to avoid the cryptocurrency market's volatility. Cryptocurrency payment has several advantages, including improved transactional security, fraud protection, a decentralized framework, low fees, customer chargeback protection, and quick international transfers.

Cryptocurrencies which are true use blockchain technology to ensure that transactions are correctly recorded, which makes hacking extremely difficult. Blockchain software is a decentralized ledger that is not managed by a single person or institution; since all transactions are registered through multiple nodes, transparency is provided. It is extremely difficult for any user to manipulate the system.

The fundamental issue with all cryptocurrencies is their fluctuating value. While holding any specific digital currency, you may lose (or gain) money until you use it to buy something. Modern money (referred to as fiat currency by cryptocurrency supporters) appears to be more robust, backed by governments and a much larger global trading network.

So, if you are a businessman or an entrepreneur looking to get their feet wet and explore this new world, or even if you are a curious minded person who just wants to explore the concepts of Cryptocurrency, then you are in the right place!

Being a relatively new form of currency, thousands and thousands of people are still confused about what Cryptocurrency is. Many individuals are interested in penetrating into this world and pursuing this literal goldmine. Still, they are unable to do it due to a lack of knowledge and proper resources. With this book, I aim to break that barrier and make it easier for absolute beginners to dive into the world of Cryptocurrency and start trading right away!

Throughout this book, we will be going through the very basics of cryptocurrency and covering its fundamental concepts. After that, we will explore the different types of Cryptocurrency, the advantage and disadvantage of investing, and give you step-by-step guides on how you can start developing your own Cryptocurrency or start trading/investing in them.

Since my target audiences are absolute beginners, I tried my very best to break everything into smaller chunks so that you can easily absorb the information at your own pace.

If you stick with me until the end, I promise you that you will find something of value within the pages of the book.

Let's move to the first chapter now, where I explain the fundamentals of Cryptocurrency and start building up on the concept from there.

Chapter 1

All You Need To Know About Cryptocurrency

What Is Cryptocurrency?

Cryptocurrencies, also known as virtual currencies, are digital currencies which are created and used by private individuals or groups.

Because most cryptocurrencies are not regulated by national governments, they are classified as alternative currencies — mediums of financial exchange that exist outside the scope of state monetary policy.

Bitcoin (BTC) is the most prominent cryptocurrency and the first to be widely used. However, hundreds of cryptocurrencies exist, and new ones emerge every month.

To distinguish them from the original, non-Bitcoin cryptocurrencies are referred to as "altcoins."

Most cryptocurrencies are functional variants of Bitcoin which is the first widely used cryptocurrency.

Cryptocurrencies, like traditional currencies, express value in units — for example, you can say, "I have 2.5 Bitcoin." just as you would say, "I have $2.50."

Cryptocurrency users benefit from advantages which are not available to traditional fiat currencies, such as the US dollar, and have the financial systems that those currencies support due to their political independence and essentially impenetrable data security.

For example, while a government can easily freeze or even seize funds held in a bank account located in its jurisdiction, it is much more difficult to do so with funds held in cryptocurrency — even if the holder is a citizen or legal resident.

On the other hand, cryptocurrencies have several risks and drawbacks that many fiat currencies do not have, such as illiquidity and value volatility.

Furthermore, because cryptocurrencies are frequently used to facilitate gray and black-market transactions, many countries regard them with suspicion or outright hostility.

While proponents tout cryptocurrencies as potentially profitable alternative investments, few serious financial professionals regard most coins — with the notable exception of Bitcoin and a few others — as suitable for anything other than pure speculation.

Advice: Are you thinking about investing in Bitcoin or another cryptocurrency? Coinbase is one of the most popular platforms for buying and selling Bitcoin and other cryptocurrencies.

The Worth Of Cryptocurrency

According to CoinMarketCap.com, a market research website, over 10,000 different cryptocurrencies are publicly traded. And cryptocurrencies continue to proliferate, with initial coin offerings or ICOs which are used to raise funds. According to CoinMarketCap, the total value of all cryptocurrencies on May 27, 2021, was more than $1.7 trillion, a decrease from the April high of $2.2 trillion. The total value of all bitcoins, the most popular digital currency, was estimated to be around $735 billion, down from a peak of $1.2 trillion in April.

Cryptocurrency	Market Capitalization
Bitcoin	$735.5 billion
Ethereum	$324.2 billion
Tether	$61 billion
Binance Coin	$57.5 billion
Cardano	$54.6 billion
XRP	$46.5 billion
Dogecoin	$44 billion
Polkadot	$22.1 billion
USD Coin	$21.9 billion
Internet Computer	$16.7 billion

Understanding How Cryptocurrency Works

It is true that the source code and technical controls which supports and secures cryptocurrencies are extremely complex. However, laypeople are more than capable of grasping the fundamental concepts to become knowledgeable cryptocurrency users.

Several concepts govern the values, security, and integrity of cryptocurrencies.

Cryptography

Cryptocurrencies secure their units of exchange through cryptographic protocols, which are extremely complex code systems that encrypt sensitive data transfers.

Cryptocurrency developers create these protocols using advanced mathematics and computer engineering principles, making them nearly impossible to break and duplicate or counterfeit the protected currencies.

These protocols also conceal the identity of cryptocurrency users', making transactions and fund flows difficult to attribute to specific individuals or groups.

Blockchain Technology

The blockchain of a cryptocurrency is the master public ledger that records and stores all previous transactions and activity, validating the ownership of all currency units at any given time.

A blockchain is the record of a cryptocurrency's entire transaction history to date. It has a finite length — containing a finite number of transactions — that grows over time.

Every node of the cryptocurrency's software network stores identical copies of the blockchain — the network of decentralized server farms run by computer-savvy individuals or groups of individuals known as miners, constantly record and authenticate cryptocurrency transactions.

A cryptocurrency transaction isn't technically complete until it's added to the blockchain, which is usually within minutes. When a transaction is completed, it is usually irreversible.

Unlike traditional payment processors such as PayPal and credit cards, most cryptocurrencies do not include built-in refund or chargeback functions, though some newer cryptocurrencies do.

The units are not available for use by either party during the lag time between the transaction's inception and completion. They are instead held in escrow — a limbo, for all intents and purposes.

Thus, the blockchain prevents double-spending or manipulating cryptocurrency code to duplicate and it sends the same currency units to multiple recipients.

Decentralized Control

The principle of decentralized control is inherent in blockchain technology.

The supply and value of cryptocurrencies are determined by the actions of their users and the highly complex protocols built into their governing codes rather than by conscious decisions made by central banks or other regulatory authorities.

The activities of miners, in particular, are critical to the stability and smooth operation of currencies. Miners are cryptocurrency users who use vast amounts of computing power to record transactions in exchange for newly created cryptocurrency units and transaction fees paid by other users.

Private Keys

Every cryptocurrency owner has a private key that allows them to exchange units and authenticate their identity. Users can generate their own private keys, which are formatted as whole numbers up to 78 digits or a random number generator can be used to generate private keys.

They can obtain and spend cryptocurrency once they have a key. Without the key, the holder cannot spend or convert their cryptocurrency, effectively rendering their holdings worthless until the key is recovered.

While this is an important security feature that reduces theft and unauthorized use, it is also extremely harsh.

Losing your private key is analogous to tossing a wad of cash into a garbage incinerator.

Although you can generate a new private key and begin accumulating cryptocurrency again, you will be unable to recover the holdings secured by your old, lost key.

Savvy cryptocurrency users are thus maniacally protective of their private keys, typically storing them in multiple digital locations — which are not Internet-connected for security reasons — as well as on paper or in another physical form.

Cryptocurrency Wallets

Cryptocurrency users have wallets that contain unique information that identifies them as the owners of their units.

Wallets reduce the risk of theft for units that aren't being used. In contrast, private keys confirm the authenticity of a cryptocurrency transaction.

Cryptocurrency exchange wallets are somewhat vulnerable to hacking. For example, the Japan-based Bitcoin exchange Mt. Gox declared bankruptcy a few years ago after hackers systematically stole more than $450 million in Bitcoin exchanged on its servers.

Wallets can be kept in the cloud, on a hard drive, or on an external storage device. It is strongly advised to keep at least one backup of a wallet, regardless of how it is stored.

It is important to note that backing up a wallet does not duplicate the cryptocurrency units themselves, it only records their existence and current ownership.

Miners

Miners keep records for cryptocurrency communities and act as indirect arbiters of the currencies' value.

Miners use highly technical methods to verify the completeness, accuracy, and security of currencies' blockchains, often manifested in private server farms owned by mining collectives comprised of dozens of individuals.

The scope of the operation is similar to the search for new prime numbers, which also necessitates massive amounts of computing power.

Miners' work creates new blockchain copies regularly, including recent, previously unverified transactions that haven't been included in any previous blockchain copy — effectively completing those transactions.

Each addition is referred to as a block. Blocks are made up of all transactions since the last new copy of the blockchain was created.

The term "miners" refers to the fact that miners' literally create wealth in the form of new cryptocurrency units.

In fact, each newly created blockchain copy comes with a two-part monetary reward: a fixed number of newly minted ("mined") cryptocurrency units and a variable number of existing units collected from optional transaction fees paid by buyers (typically less than 1% of the transaction value).

It's worth noting that cryptocurrency mining was once a potentially lucrative side business for those with the means to invest in power- and hardware-intensive mining operations.

Today, it is impractical for hobbyists to invest in professional-grade mining equipment unless they have thousands of dollars to spare. If you just want to supplement your regular income, there are plenty of freelance opportunities that will pay you more.

Although sellers are not charged transaction fees, miners can prioritize fee-loaded transactions over fee-free transactions while creating new blocks, even if the fee-free transactions came first in time.

Because sellers have an incentive to charge transaction fees because they get paid faster, it is fairly common for cryptocurrency transactions to include fees.

Although it is theoretically possible for previously unverified transactions on a new blockchain copy to be fee-free, this rarely occurs in practice.

Cryptocurrencies automatically adjust to the amount of mining power working to create new blockchain copies via instructions in their source code — copies become more difficult to create when the mining power increases and it becomes easier to create when the mining power decreases.

The goal is to maintain a constant average interval between new blockchain creations. For example, Bitcoin's it is usually 10 minutes.

Finite Supply

Although mining generates new cryptocurrency units regularly, most cryptocurrencies are designed to have a limited supply — a key guarantor of value.

In general, this means that miners will receive fewer new units per new block as time passes. Miners will eventually only be paid transaction fees for their efforts. However, this is yet to happen in practice and may not happen for some time in future.

If current trends continue, the last Bitcoin unit might be mined sometime in the mid-22nd century, which is still a long way to go.

The finite supply of cryptocurrencies makes them inherently deflationary, making it more akin to gold and other precious metals, which have finite supplies, than fiat currencies, which central banks can, in theory, produce unlimited supplies of.

Cryptocurrency Exchanges

Many lesser-used cryptocurrencies can only be exchanged via private, peer-to-peer transfers, making them less liquid and difficult to value compared to other currencies, crypto, and fiat.

More popular cryptocurrencies, such as Bitcoin and Ripple, are traded on secondary exchanges similar to forex exchanges for fiat currencies. (One example of an exchange is the now-defunct Mt. Gox.)

These platforms enable cryptocurrency holders to exchange their holdings for major fiat currencies such as the US dollar, euro and other cryptocurrencies, including less popular currencies.

In exchange for their services, they take a small percentage of each transaction's value — usually less than 1%.

Importantly, cryptocurrencies can be exchanged for fiat currencies in special online markets, which means that each cryptocurrency has a variable exchange rate with major world currencies like the US dollar, British pound, European euro, and Japanese yen.

Cryptocurrency exchanges play an important role in establishing liquid markets for popular cryptocurrencies and determining their value over traditional currencies. You can even trade cryptocurrency derivatives or track broad-based cryptocurrency portfolios in crypto indexes on some cryptocurrency exchanges.

However, exchange rates can remain extremely volatile. For example, in the aftermath of Mt. Gox's collapse, Bitcoin's US dollar exchange rate fell by more than half, it had increased more than tenfold in 2017 when the cryptocurrency demand exploded.

And cryptocurrency exchanges are somewhat vulnerable to hacking, serving as the most common location for digital currency theft by hackers and cybercriminals similar to those responsible for the takedown of Mt. Gox.

The History Of Cryptocurrency

Bitcoin is the world's most popular cryptocurrency, thanks to it being the first to ever exist. The coin was created just as the world was emerging from the depths of the economic crisis that affected the housing and stock markets. Much of Bitcoin's initial appeal centered around the state of the financial system at that time.

It's worthwhile examining the backdrop in which Bitcoin was released. Beginning in 2007, the world began to realize that investment banks and lenders had been running major risks in the housing markets. Lenders were unscrupulously underwriting loans to people who couldn't afford them using false advertising and less than ethical promises.

They were doing this because Wall Street was buying all of these loans, repackaging them into complex securities, and selling them to investors as a safe source of income. The problem was that as lending standards deteriorated, the security of these instruments dropped, and they became time bombs.

Wall Street, despite this, did not play it safe and instead borrowed vast sums of money and began buying these securities themselves. As a result, when they started going bad, Wall Street's losses exceeded compared to what they had invested. It began with a few hedge funds toppling, and then investment banks began going under as well.

The government's response was to use taxpayer's money to bail out these firms. The idea was to use this money to buy the toxic assets these banks held to prop them up. Led by the Treasury Department and the Federal Reserve Bank, the government bought billions of these assets as they sought to avoid economic collapse.

However, these actions were hollow because the only entities that were bailed out were investment banks. Ordinary people received nothing and were left dealing with foreclosures. While big banks borrowed money to bail them out of poor decisions, "Main Street" was told that they'd have to live with the consequences of being sold bad loans.

The rigorous manner in which people like Hank Paulson (the Treasury Secretary at that time) patted themselves on the back for their actions but it also rubbed people in the wrong way. The impression that the government was catering only to the rich was proved true. Not a single CEO went to jail or faced any consequences. No lenders were prosecuted. The rating agencies that had committed outright fraud against investors were left alone.

To prop up the economy, the Federal Reserve began printing paper money, hoping that an increased money supply would drive spending once again and energize business. Again, the mistake here was that this money was given to banks. These banks were then told to follow strict lending criteria.

As a result, people who had recently suffered losses and therefore needed money the most were ineligible to borrow money. It went instead to people who needed it the least, the ones who hadn't suffered during the downturn. These people received cheap debt to boost their net worth, and the banks made money. All the while, the majority that needed money received no assistance. With this backdrop, it was obvious that government and big finance had conspired to create an unequal society. Centralization no longer made sense, and it was time to create an alternative system of finance.

ENTER BITCOIN

On October 31, 2008, as the world was still reeling from the collapse of major banks, a person name Satoshi Nakomoto distributed a message on his crypto mailing list. This message contained a link to a whitepaper entitled "Bitcoin: A peer-to-peer electronic cash system." The boring title belied the immense nature of what was being proposed.

Within this paper, Nakomoto outlined his vision for a fully decentralized currency that would operate outside of the world's financial system. There wouldn't be a single administrator who would control the supply of coins. Instead, the transaction ledger would be stored on nodes or computers around the world. These nodes would belong to people who owned the coins.

Miners, who would be paid to maintain the network and create new coins, would say how the network ran and point out any developments that might be necessary. Thus, everyone on the network would act as a collective central bank. This currency would be housed on a blockchain network, which guaranteed privacy and security.

INITIAL REACTION

Given the highly technical nature of the subject and the esoteric nature of what was being proposed, Bitcoin's initial announcement went unnoticed. Most people were instead focusing on the events that were playing out politically. As outrage over government action grew, more people began noticing Bitcoin.

Back then, there were no exchanges, and almost nobody knew what a cryptocurrency even was. The extensive literature that one can access today for free wasn't around, nor there were any large forums of people discussing the topic. It took people a long time to even understand what was being proposed.

Mainstream media covered Bitcoin being an alternative currency, which wasn't entirely true. Yes, Bitcoin was a currency, but its real aim was to create a separate financial network entirely. One of the things that interested mainstream viewers the most was that no one knew who Nakamoto was. To this day, this person's or people's identities are a secret.

Bitcoin was being traded in electronic networks by people who understood how the technology worked. The prospect of owning crypto wallets, keys, and so on confused most people. Therefore, the bulk of the initial Bitcoin mined was concentrated in the hands of the few.

2010 saw the first real-world transaction in Bitcoin. An unknown person paid for two pizzas with 10,000 Bitcoin.

It's safe to say that these two pizzas are probably the most expensive in the history of the world, given Bitcoin's current valuation. However, back then, no one knew what to make of it.

For the most part, Bitcoin was viewed as a joke and a symbolic resistance to corporate greed. No one knew where it would go or what kind of financial system it would lead to. As 2010 turned to 2011 and America began emerging from its financial crisis, social inequality started surfacing.

This led to increased interest in cryptocurrencies, and as a result, rival coins began popping up.

RIVAL OPTIONS

The first rival to Bitcoin to emerge was Ether (Ether), based on the Ethereum blockchain. Given that all of these currencies were lines of code, anyone could create them. This led to a wave of people creating their own cryptos, and as a result, the amount of fraud began increasing. For instance, people would create a coin (say null coin), publish its price as a few fractions of a cent, and then rely on euphoria to boost prices to one cent or more.

Having made many multiples of their money, they would then exit the coin and leave others holding the bag. This isn't to say that all of these so-called altcoins were fraudulent. Coins such as Ether and Litecoin had significant technical investments behind them and were created with specific aims.

The general public wasn't used to reading technical whitepapers, which resulted in them investing their money into everything that sounded like a good idea. The number of altcoins that popped up numbered well above 1,000. Simultaneously, initial crypto trading exchanges began appearing online, along with wallet storage services.

All crypto holdings need a secure wallet, and this created a secondary business opportunity. However, fraud was prevalent even here. Many online services that posed as wallets were fraudulent schemes designed to steal people's Bitcoin holdings. The boom in altcoins led to the price of Bitcoin surge.

It breached the $1,000 mark for the first time in 2013, at which point it promptly crashed to $300. It would be another two years before it would breach $1,000 again. This massive price rise created questions on its viability as a means of transaction. Most of these questions remain unanswered in the euphoria surrounding cryptocurrency. How can something whose price varies as much as Bitcoin's be used as everyday currency?

Imagine paying for something in the morning and then discovering that your coins are now worth half their value in the evening. This is exactly like the hyperinflation scenario in Venezuela that was highlighted in the previous chapter. Regardless of this, it was clear that speculative hype was pushing prices higher. Volatility was ingrained into Bitcoin prices at this point as prices rose and fell by over 20% daily.

Trading in Bitcoin was restricted to offline exchanges. People would physically meet up and exchange coins or on online exchanges such as Mt. Gox. Gox was the biggest crypto exchange at that time and facilitated trading in several cryptocurrencies. Many people invested their cash into Bitcoin at $1,000 price only to see it plummet. This would be a recurring experience for Bitcoin speculators.

SCAMS

As the trading volume in Bitcoin grew, so did the number of fraudulent operations. The first operation that came to light was the Dark Web online marketplace called Silk Road. This website was a marketplace that sold everything from drugs to stolen credit card information. The U.S. government arrested everyone involved and shut down operations in 2013.

However, there were many other kinds of fraud that could not be regulated. One example of this was creating bogus cryptocurrencies that had no value or technical intent behind them. Thefts from online wallets were now normal, and soon the biggest exchange, Mt. Gox, was also targeted.

The Mt. Gox incident remains the largest instance of Bitcoin theft. In January 2014, the exchange went offline, and it claimed that over 850,000 Bitcoin had disappeared. Due to heavy trading volumes, most exchange users stored their Bitcoin on the exchange's wallet. At that time, this haul was worth $450 million, which made it a significant heist. The present value of the lost coins would be around $4.4 billion.

The Mt. Gox incident was one of the first instances of crypto fraud hitting mainstream users instead of Dark Web actors. For the first time, regular users suffered due to security lapses on the exchange. Needless to say, Mt. Gox never recovered from that incident. As a result, greater recognition was given to the need for security.

Wallet best practice and storage needs began to be addressed with greater urgency. The high price of Bitcoin resulted in people recognizing that the currency wasn't a joke anymore. It was a real asset which was worth a lot of money, and it needed to be secured. The result was a booming market in wallet storage as crypto wallets became more and more sophisticated.

THE INITIAL COIN OFFERING WAVE

As Bitcoin's prices grew, so did the value of Ether and other cryptos. This led to several new coins being offered, and creative financing methods began cropping up. One such venue that was often abused was the 'ICO,' or initial coin offering. Much like how stocks are offered to the public for the first time via an Initial Public Offering, an ICO was the first to issue a new coin.

ICOs began cropping up everywhere globally, and needless to say, there were quite a few fraudulent currencies involved. Some ICOs

offered investors steady monthly payments while others promised to pay investors back after their money was invested in physical businesses.

In short, the ICO became a way for many people to raise money, legitimately or illegitimately. For instance, instead of raising money by approaching investors, a person could launch an ICO for a new coin. This coin's value would be tied to the value of the investments which the proceeds from the ICO would be put toward.

Many people invested in ICOs, and quite a few of them lost money. The problem was that the average person was still unwilling to take the time to read whitepapers and understand what these coins were about. It was almost inevitable they would lose money as a result.

The U.S. government warned the public that these ICOs were unregulated and might be Ponzi schemes. The Chinese government went a step further and banned them outright. This didn't deter scammers, though. They simply took their business to offshore tax-havens.

A MAJOR MILESTONE

As the decade unfolded, it became increasingly clear that Bitcoin and cryptos as an asset class were becoming relevant. Major political events caused the prices of Bitcoin to rise higher than ever. For instance, the Indian government's ambitious scheme to declare certain forms of cash illegal sent Bitcoin prices soaring.

The instability of the economies of Venezuela and Zimbabwe offered people more use cases for Bitcoin. More and more everyday purchases could be paid using Bitcoin. Many merchants started accepting Bitcoin as payment for goods and services.

Meanwhile, the blockchain itself was gaining widespread traction as a technological solution. Digital transformation was increasing as a business priority, and this meant that cybersecurity had moved from something IT departments dealt with to becoming a concern for the entire business. Blockchain and its inherent security features offered a good solution to this issue.

Institutional finance began taking notice of cryptocurrencies as a new wave of Silicon Valley-led entrepreneurs lent their approval to these alternative forms of finance. The result was that by 2017, Bitcoin finally breached the $10,000 threshold. Once again, it promptly crashed back down as it had previously. However, by this point in time, the waves in which Bitcoin moved were obvious to most people, and they continued to hold onto their coins.

By the middle of 2018, Bitcoin had cleared $20,000, and there was a lot of talk of a price bubble and the possibility of a huge crash. While prices did crash shortly after that, they still managed to remain above previous lows. In fact, the crashes took prices close to previous highs.

As 2019 came to an end, Bitcoin prices were hovering around the $15,000 mark, and they had remained stable for a long time. This was about to change.

THE PANDEMIC

The COVID-19 pandemic took some time to gain a foothold in everyone's consciousness. What began as an outbreak in China's Wuhan province managed to become a global pandemic that cost millions of lives. Thanks to poor government handling and many false or uneducated statements at the start of the outbreak. As more examples of the lies that centralized systems propagate in society unfolded, it became clearer that decentralization was the answer.

Bitcoin prices rose slowly at first. However, mainstream financial companies began recognizing the viability of Bitcoin as an investment and began moving client funds into it. Paypal announced it would facilitate Bitcoin payments, and many investment banks announced wealth management options in Bitcoin.

A Bitcoin ETF was launched in Canada, which made it even more accessible for ordinary investors. As 2020 unfolded, the price of Bitcoin rose to eye-watering heights. It eventually passed $50,000, and as 2021 began, its price was sitting close to $60,000. People who had bought

Bitcoin when it was first issued and had held on were now billionaires.

There was renewed interest in Nakamoto's real identity at this point. Still, it was clear that this person or entity was focused on remaining anonymous. Their wallet address is public knowledge, and records show that they haven't sold any of their holdings since the beginning.

Nakamoto's wallet remains a source of great interest since any activity could lead to major price fluctuations. Thus far, the wallet hasn't conducted any major transactions.

Throughout this price rise, there's been a lot of debate about whether Bitcoin is a good investment or not; but it is clear that the value of Bitcoin is divorced from its price. There's no doubt that Bitcoin can be used as an alternative currency, but how much should it be worth? No one knows for sure.

It's a tough asset to value since it has a limited supply. Yet a lot of its price is derived from what others think about it. Therefore, even institutional managers can't get a handle on how much they ought to pay for it. Either way, there's no denying that Bitcoin is here to stay, as is the rest of the crypto world.

The prices of altcoins have also risen significantly throughout the pandemic. Along the way, even troll currencies such as Dogecoin have witnessed increasing prices, which shows the sometimes-worrying levels of ignorance that many market participants possess. However, most cryptocurrencies have real-world applications and have different aims.

For instance, Monero aims to provide extreme levels of privacy to its users. While it might not prove to be a viable everyday mode of transaction, there's no doubt that it has utility in certain cases. Investing in cryptocurrencies remains fraught with risk - thanks to the huge volatility that each coin experiences.

Bitcoin still rules the roost as the biggest cryptocurrency around. Ether is also popular with investors, but its aims are broader. In addition to hosting the Ether currency, Ethereum's blockchain is positioned as the network of choice for many decentralized financial applications. This makes its reach more ambitious than that of any other cryptocurrency network out there.

As for Bitcoin, it continues to be a haven for people looking for an asset that is non-correlated with the rest of the market. The stock market was crushed during the initial stages of the pandemic, but Bitcoin managed to rise by

224% in the first half of 2020. This sort of behavior has ultimately convinced mainstream finance to move into Bitcoin.

Understand that this doesn't mean Bitcoin isn't risky anymore. There will always be risks involved when investing in any kind of asset.

However, as time progresses, it's become clear that many of the perceived risks in Bitcoin aren't real and have many real-world applications.

The Future Of Cryptocurrency

People used to trade goods and services directly through a system of bartering. Currency evolved over time to speed up and standardize the transactions. The first use of an industrial facility to produce coins for use as a currency is thought to have occurred around 600 B.C. when the elites of Lydia (modern-day Western Turkey) used stamped silver and gold coins to pay armies. We've come to rely on fiat currencies (coins, banknotes, and credit cards) to buy goods and services over the centuries. The twenty-first century has brought us yet another currency: virtual currency, also known as cryptocurrency. In response to the 2008 financial collapse, Satoshi Nakamoto created Bitcoin, the first cryptocurrency to hit the market in 2009.

There are currently over 4,000 different types of cryptocurrencies circulating in the global digital infrastructure. These virtual currencies can be used to purchase goods and services by utilizing an online ledger with strong cryptography to secure online transactions built on blockchain technology.

Getting into Blockchain

Blockchain can be quite complex but consider it like a database that stores information in blocks, with chains connecting the blocks in chronological order. A blockchain can store various types of information; for example, it is used in a land registry to record structural changes to buildings, and "if the property is sold, all relevant documentation can be transferred to the new owner." Every transaction is auditable, timestamped, and irrefutable." So far, the most common application of blockchain has been as a transaction ledger. Blockchain technology is the backbone and enabler of cryptocurrency's existence.

Dissolution Of Power

Bitcoin transactions are recorded on a blockchain in a decentralized manner, which means that no single person or group has control over how data is stored, but rather all the users retain control collectively. This preserves the ledger's integrity and prevents people from corrupting the system, in contrast to our traditional currencies which are controlled by our central banks and the government. This does not imply that decentralized systems are immune to manipulation. Still, it does imply that there is no single point of attack, making them highly hack-resistant.

Because blockchain is decentralized and not controlled by a single authority, central banks and governments fear it because they cannot manipulate the value of cryptocurrency in the same way as traditional currencies. This means that if we use cryptocurrency as our primary form of currency, central banks and governments will be unable to intervene and, as a result, will be unable to control the rate of inflation. Furthermore, blockchain transactions eliminate government tax revenue and financial institution interest. While this sounds appealing, there are understandably concerns and uncertainties about the nature and applications of cryptocurrencies.

Usage Of Cryptocurrency In Transaction

Legal tender refers solely to fiat money and is accepted as satisfactory payment to discharge a debt. Even though cryptocurrencies lack traditional legal tender, there is no reason why they cannot be accepted as a form of payment. For starters, credit card payments lack legal tender but are widely accepted as a form of payment. Secondly, Bitcoin is already accepted as a form of payment by Starbucks, Whole Foods, and other major retailers. Flexa, a payments startup based in the United States, believes that "bringing cryptocurrency to the masses is the best way for global commerce to become more efficient and accessible."

Furthermore, Elon Musk's electric car company, Tesla, purchased $1.5 billion (£1.1 billion) in Bitcoin this year. It stated that it intends to begin accepting it as payment shortly. This purchase resulted in a rise in Bitcoin's value, demonstrating that an increase in confidence in the crypto markets equals market value.

Putting Rules Over Blockchain And Cryptocurrency

There is without a doubt a significant need to educate everyone on blockchain and cryptocurrencies to prepare for the opportunities that come with new emerging technologies. However, in the last 12 years since the introduction of cryptocurrency, it appears that the rate at which people are becoming familiar with cryptocurrency is indeed very slow. It appears that financial authorities and governments around the world are beginning to scrutinize cryptocurrencies rather than educating people on something that is already taking shape.

The UK's Financial Conduct Authority (FCA) imposed a regulatory ban on cryptocurrency trading in January 2021, demonstrating authorities' skepticism and lack of control over cryptocurrencies. Across the Atlantic, near the end of 2020, the US Securities and Exchange Commission (SEC) filed a multibillion-dollar lawsuit against Ripple Labs (Ripple) and its executives for selling XRP, claiming it does not meet the criteria of a currency and instead fits the criteria of an investment contract, and is thus security.

Near the end of 2020, US Securities Exchange Commission (SEC) filed a multi-billion dollar lawsuit against Ripple Labs (Ripple) and its executive.

They claimed that the concept of cryptocurrency trading did not fit a criteria in the investment contract, which might lead to security issues in the future, .

The SEC is now requesting injunctive relief as well as disgorgement of allegedly ill-gotten gains.

It is worth noting that the SEC considers cryptocurrencies — Bitcoin and Ethereum — to be currencies, which raises the question of what constitutes a currency and what constitutes a security in the crypto space. According to the SEC, Bitcoin is classified as a currency. It is decentralized, whereas XRP is classified as a security because it is controlled by a central authority.

Many people believe that the SEC's "misguided" lawsuit against Ripple will harm cryptocurrency even more, and this has proven to be true. Since the lawsuit was filed, various exchanges have begun to delist XRP, raising public concerns about the value of cryptocurrencies in general. On the other hand, it is hoped that this lawsuit will result in the necessary regulation which will boost certainty and public confidence. It was recently reported that on March 9, 2021, US lawmakers introduced a bipartisan bill to clarify crypto regulation. Perhaps this bill will provide the public with much-needed clarity about the differences between a currency and security, allowing them to make their own decisions.

The Volatility Of The Market

Bitcoin was worth less than a US dollar when it was first introduced. It is now worth more than $50,000 (£36,000) in the United States. Because there are only 21 million Bitcoins that can be mined, the price increase is due to simple supply and demand. As of February 24, 2021, 18.638 million Bitcoin had been mined, leaving 2.362 million to enter circulation. Decisions such as those made by the SEC and FCA add to the volatility in the cryptocurrency market, possibly unintentionally causing more uncertainty in the overall crypto market, resulting in traders making decisions based on emotion, thus increasing market volatility and quick liquidity.

To Sum Up

People became interested in cryptocurrency after the 2008 financial crisis because they lost faith in the existing financial infrastructure. Major events such as WWI, WWII, and the 2008 financial crisis had long-term and devastating effects on the global economy. Following the current COVID-19 pandemic, our livelihoods and economy have taken yet another hit amid recovery. Central banks are printing more money to relieve our financial burdens. Still, we must ask how long this process of quantitative easing can be sustained.

Our traditional currencies are losing value as a result of this process. Today, our money has value because our government tells us so. Still, the US dollar (the world's reserve currency) has not been backed by gold or anything of intrinsic value since the Nixon Shock in the 1970s. With rising living costs and a bleak outlook for our economy's recovery, it's no surprise that we're looking for a golden parachute.

Businesses and individuals alike are adapting based on the preceding and the current rate of technological advancement. Consider fintech firms, which have already begun to access new blockchain-based technologies to stay ahead of the competition. While authorities impose cryptocurrency regulations, they recognize that they cannot halt the march of progress because "markets and technology are always changing," and "fintech can be a powerful force for good" (Gensler, 2021).

This could imply that blockchain and cryptocurrency will become mainstream in the next five to ten years.

Chapter 2

The Different Types Of Cryptocurrency

The Different Of Crypto Currency

Bitcoin (BTC)

Referred to as "digital gold" or the gold standard for cryptocurrency, Bitcoin, which debuted in 2009, has reigned king over every other cryptocurrency.

With a market capitalization of over $900 billion, Bitcoin dominates the rest of the cryptocurrencies with the lion's share of the total cryptocurrency market cap.

Investing in just one Bitcoin is an expensive endeavor compared to other investment vehicles. Just for comparison, one bitcoin is equivalent to $48,920.30.

Litecoin (LTC)

Referred to as the "silver to Bitcoin's gold," Litecoin was created as a fork (or split) from Bitcoin and released in 2011 as competition.

Litecoin was made to process transactions faster and cheaper than Bitcoin. One Litecoin is equal to about $178.93.

Ethereum (ETH)

Ethereum is another cryptocurrency powerhouse, but it isn't meant to be a peer-to-peer payment system as Bitcoin.

Ethereum was launched in 2015 as a decentralized software platform that powers smart contracts (programmatically enforced contracts) and distributed applications ("decentralized" apps or dApps).

dApps

These decentralized applications are open source, autonomous, have 100% uptime, and leverage a blockchain's benefits (no central server, extremely difficult to be hacked, etc.)

Imagine the apps in the iTunes store are not of their own entities iand are not centralized through Apple — that's what a dApp is. Nearly 3,000 dapps use Ethereum's blockchain .

Smart Contracts

These are strings of code that automatically execute a certain task when specific conditions are met. For example, Alex could set up a smart contract to "pay Steven $40 if he sends 10 unique logo designs by December 8th, 2021."

Once Steven completes this task, the smart contract automatically pays him the $40. If he doesn't, then Alex is returned his $40.

The Types Of Cryptocurrency

Let's go over a small difference that many people get tripped over when understanding cryptocurrency basics. There are two different types of cryptocurrency: coins and tokens.

Coins

A coin is a cryptocurrency with its own blockchain, such as Bitcoin, Ethereum, Litecoin, Ripple.

When someone says they "bought cryptocurrency," they are referring to buying coins.

Token

A token is a cryptocurrency built on another blockchain, for example, a dApp that runs on Ethereum's blockchain.

Tokens represent an asset or utility for a specific project. They are sold (or given) during the first public sale for a project, an Initial Coin Offering (ICO), which mirrors an Initial Public Offering in the stock market.

There is another very important distinction with tokens. There are two general types of tokens: utility and security.

Utility Token

A utility token is intended only to buy products or services from the company or platform that issues them. A security token is essentially a digital version of financial security that acts as a share of the value of an enterprise, like owning AAPL which essentially means you own a chunk of Apple.

In other words, security tokens pay dividends, share profits, pay interest or invest in other tokens or assets to generate profits for the token holders. A digital asset is considered a security token if it meets three criteria: It requires a monetary investment. The collected funding goes to a single enterprise. Investors give their money with the expectation of gaining income derived from the work of the third party. Security tokens must also be fully compliant and follow these regulations:

Regulation D: The individual offering in the security can only raise money from accredited investors. The information provided to them is "Free from false or misleading statements" (Section 506C).

Regulation A+: An exemption that allows the creator to solicit non-accredited investors with SEC-approved security for up to $50 million in investment. This option takes a lot more time and is generally the most expensive route for issuance.

Regulation S: This regulation outlines security offerings from countries outside of the US, which are therefore not subject to the registration requirements of section 5 of the 1993 Act. The creators of the security offering still must follow the security regulations of the country that they plan to solicit investment.

The Best Cryptocurrencies To Consider

The most daring investors can purchase new cryptocurrencies at the initial coin offering or shortly after that, following the example set by many of today's Bitcoin millionaires. However, there may never be a "next Bitcoin." The estimated 1,000 cryptocurrencies that have vanished into the digital ether point to risks for early investors. Let us have a look at which coins made the list.

Bitcoin (BTC)

The granddaddy of all cryptocurrencies, Bitcoin was first and is the most well-known cryptocurrency on the market. It also benefits from the largest market cap. It is among the most highly traded cryptocurrency, assuring liquidity in the short term. Bitcoin is the king for retail adoption, leading all other cryptocurrencies in terms of acceptance as a payment medium.

Down significantly from its all-time high of over $20,000 per Bitcoin, BTC may have plenty of room for growth despite an increasingly crowded field of competitors.

Ethereum (ETH)

As the currency and platform made "smart contracts" as part of the cryptocurrency market's vocabulary, Ethereum has seen massive gains since its introduction in 2015. Currently trailing behind Bitcoin regarding market capitalization, Ethereum has become one of the most widely discussed cryptocurrency projects globally.

A consortium of some of the biggest names in the business, including Microsoft, Intel, Chase, and J.P. Morgan, are building business-ready versions of the software that drives Ethereum. With momentum and market enthusiasm behind the Ethereum project, there's no reason to think Ethereum has run its course, and investors should consider Ethereum as part of a cryptocurrency portfolio.

Ripple (XRP)

Ripple diverges from much of its cryptocurrency competitors in several ways. Ripple is an invention of Ripple Labs, and the Ripple token is being used in high-speed, and low-cost money transfers worldwide. Ripple Labs has announced several partnerships with leading money transfer services, with more financial market partnerships expected in the future.

Unlike many cryptocurrencies that trade on hopes and dreams, Ripple is being used in the real world today, showing signs of future adoption within the financial market community. Ripple rose in value over 36,000% in 2017, but similar gains may not be likely going forward.

EOS (EOS)

Another cryptocurrency with smart contracts like Ethereum, which is gaining in popularity, is EOS. EOS is credited with being the first blockchain operating system, offering decentralized applications that live on the blockchain and parallel processing, enabling faster transaction speeds and better scalability than some competitors. Transactions on the EOS network is free.

Many competitors, including Ethereum, have a transaction fee for transferring coins or tokens from one wallet address to another. EOS concluded its year long ICO in May this year, raising a total of $4 billion. The longer-duration ICO was done to create an orderly market for EOS without the dramatic run-up and sudden crash which is common to cryptocurrencies when launched.

YTD performance for EOS is flat, with less volatility compared to some of its competitors. Enthusiasm for the project remains high, and EOS is one of the most actively traded cryptocurrencies on exchanges.

Bitcoin Cash (BCH)

Bitcoin Cash, a fork of the original Bitcoin project, is one to watch, as it's the fourth-largest cryptocurrency by market cap. If forced to level criticism against Bitcoin in its current form, slow transaction speeds are among its primary challenges.

Bitcoin cash was developed using modified code from the Bitcoin project that allowed larger block sizes, promoting faster transaction times and better scalability. Although not widely accepted or known as compared to Bitcoin, Bitcoin Cash is still a promising alternative to Bitcoin with an enthusiastic market following.

Litecoin (LTC)

Now accepted as a payment method at Overstock.com, Litecoin may also have a bright future. Long-term investors in LTC have been rewarded with up to 20x returns, although a spike in late December 2017 sent the price of LTC to over $350. Litecoin now trades at around USD 84. Although showing signs of consolidation, it remains actively traded and is consistently one of the top 10 cryptocurrencies when measured by market cap. Litecoin boasts a faster transaction time than Bitcoin, largely attributed to its use of a different algorithm to add transactions to the blockchain. Increased transaction speed also enhances scalability.

A Deeper Look Into Bitcoin

What Is Bitcoin?

Bitcoin is a digital currency that was first introduced in January 2009. It is based on the ideas presented in a whitepaper by the enigmatic and pseudonymous Satoshi Nakamoto. The identity of the person or people who invented the technology is still unknown. Bitcoin promises lower transaction fees than traditional online payment methods. Unlike government-issued currencies, they are run by a decentralized authority.

Bitcoin is a type of monetary unit. There are no physical bitcoins; only balances are kept on a public ledger which everyone can see. A massive amount of computing power is used to verify all bitcoin transactions. Bitcoins are not issued or backed by any banks or governments, and they are not valuable as commodities. Even though it is not legal tender, Bitcoin is extremely popular and has sparked hundreds of other cryptocurrencies known as altcoins. The term "Bitcoin" is commonly abbreviated as "BTC."

Understanding Bitcoin

The bitcoin system is comprised of a network of computers (also known as "nodes" or "miners") that runs bitcoin's code which is stored in blockchain. A blockchain can be viewed metaphorically as a collection of blocks. Each block contains a set of transactions. No one can cheat the system because all computers running the blockchain have the same list of blocks and transactions. One can see these new blocks filled with new bitcoin transactions in real-time.

Anyone, whether they run a bitcoin "node," can see these transactions in real-time. A bad actor would need to control 51 percent of the computing power that makes up bitcoin to commit a heinous act. As of January 2021,

Bitcoin had approximately 12,000 nodes, and this number is growing, making such an attack extremely unlikely.

However, if an attack were to occur, bitcoin miners—people who participate in the bitcoin network using their computer—would most likely fork to a new blockchain, rendering the bad actor's effort to carry out the attack futile.

Bitcoin token balances are kept using public and private "keys," which are long strings of numbers and letters linked by the mathematical encryption algorithm that created them. The public key (like a bank account number) serves as the address that is made public, to which others can send bitcoins.

The private key (like an ATM PIN) is meant to be kept private and is only used to authorize bitcoin transactions. Bitcoin keys are not to be confused with a bitcoin wallet, a physical or digital device that facilitates bitcoin trading and allows users to track coin ownership. The term "wallet" is somewhat misleading because bitcoin's decentralized nature means that it is never stored "in" a wallet but rather decentralized on a blockchain.

P2P Technology

Bitcoin was one of the first digital currencies to make use of peer-to-peer technology to enable instant payments. Bitcoin "miners" are independent individuals and businesses who own the governing computing power and participate in the bitcoin network. They oversee processing transactions on the blockchain. They are motivated by rewards (the release of new bitcoin) and transaction fees paid in bitcoin.

These miners can be thought of as the decentralized authority enforcing the bitcoin network's credibility. New bitcoin is distributed to miners at a fixed but decreasing rate. There are only 21 million bitcoins available for mining. There is approximately 18,614,806 bitcoins in existence as of January 30, 2021, with 2,385,193 bitcoins remaining to be mined.

In this way, bitcoin and other cryptocurrencies differ from fiat currency; in centralized banking systems, the currency is released at a rate corresponding to the growth rate of goods; this system is intended to maintain price stability. A decentralized system, such as bitcoin, determines the release rate in advance as per the algorithm.

Bitcoin Mining

Bitcoin mining is the process of releasing bitcoins into circulation. Mining, in general, necessitates the solution of computationally difficult puzzles to discover a new block, which is then added to the blockchain.

Bitcoin mining is the process of adding and verifying transaction records across the network. Miners are rewarded with a few bitcoins for adding blocks to the blockchain; the reward is halved every 210,000 blocks. In 2009, the block reward was 50 new bitcoins. The third halving occurred on May 11th, 2020, reducing the reward for each block discovery to 6.25 bitcoins.

To mine bitcoin, a variety of hardware can be used. Some, however, provide greater rewards than others. Certain computer chips, known as Application-Specific Integrated Circuits (ASIC), and more advanced processing units, such as Graphic Processing Units (GPUs), can achieve greater rewards. These complex mining processors are referred to as "mining rigs."

One bitcoin can be divided into eight decimal places (100 millionths of a bitcoin). The smallest unit is known as a Satoshi.

Bitcoin Investment

Many bitcoin supporters believe that digital currency is the way of the future. Many supporters of bitcoin believe that it enables a much faster, low-fee payment system for global transactions. Bitcoin can be exchanged for traditional currencies, even though it is not backed by any government or central bank; in fact, its exchange rate against the dollar attracts potential investors and traders who are interested in currency plays. Indeed, one of the primary reasons for the rise of digital currencies such as bitcoin is their ability to serve as a substitute for national fiat money and traditional commodities such as gold.

The IRS announced in March 2014 that all virtual currencies, including bitcoins, would be taxed as property rather than currency. Gains or losses on bitcoins held as capital are realized as capital gains or losses. In contrast, gains or losses on bitcoins held as inventory are realized as ordinary gains or losses. Transactions that can be taxed include the sale of bitcoins that you mined or purchased from another party, as well as the use of bitcoins to pay for goods or services.

Bitcoins, like any other asset, follow the principle of buying low and selling high. The most common way to acquire the currency is to purchase it on a bitcoin exchange, but there are numerous other ways to earn and own bitcoins.

Exploring Altcoins

By this point, it is safe to say that bitcoin has had a massive impact on the world. As a result of bitcoin's widespread popularity, many people have attempted to follow Satoshi Nakamoto's footsteps and create their own cryptocurrency. Today, there are thousands of different cryptocurrencies, effectively making bitcoin the crypto world's catalyst. These post-Bitcoin coins are known as "altcoins."

What Are Altcoins?

Following bitcoin's massive popularity and success, people began to use the Bitcoin model to create their own digital currencies.

Altcoin, the term coined for cryptocurrencies after the success of bitcoin, is derived from two words: "alt" and "coin"—meaning "alternative coin." These are commonly marketed as strong alternatives to bitcoin because they address bitcoin's limitations as believed by few people. Regardless of that selling point, altcoins must compete against bitcoin, the most popular (and most valuable) cryptocurrency globally and must have a competitive advantage over bitcoin to succeed.

Many altcoins are built on the foundation of bitcoin, giving them the much-desired feature of being “peer-to-peer.” They also aim to provide a cheap and efficient means of transaction, like bitcoin.

As of early 2020, there are over 5,000 different cryptocurrencies in the market, and altcoins account for 34% of the total cryptocurrency market, according to CoinMarketCap.

The first altcoin appeared in April 2011, just two years after the creation of bitcoin. Namecoin (NMC) was created to inject decentralization into web name registration (making user domains less visible allowing for increased censorship resistance and anonymity) and serving as an alternative digital currency to bitcoin. The coin's code was based on bitcoin's, and it even used the same proof-of-work algorithm. Namecoin, like its predecessor, has a supply limit of 21 million coins.

Litecoin was another popular altcoin that gained traction not long after bitcoin's creation. Litecoin, which debuted in October 2011, was billed as "the silver to Bitcoin's gold." Unlike bitcoin, however, Litecoin has a supply cap of 84 million, which is four times the supply of BTC. Despite following in the footsteps of Bitcoin, Litecoin was distinct in several ways, the most notable is, its approved mining and transactions much more quickly and frequently. Unlike Namecoin, which eventually died out, Litecoin is still one of the top ten cryptocurrencies in total market capitalization.

The Different Types of Altcoins

Categories within the “altcoin” umbrella have emerged as altcoins have evolved over time. Here are a few examples of the various types that have been developed over the years:

Mining Based Altcoins

These are the altcoins that arose due to bitcoin's mining process—the process by which new coins are created by solving difficult cryptographic equations to unlock blocks. Some of the most popular altcoins on the market, such as Ethereum, are mining-based.

Security Tokens

These are like traditional stocks. They frequently promise dividends, such as ownership of a business or a payout. They are frequently launched in the form of an Initial Coin Offering (ICO) and are linked to businesses.

Utility Tokens

Utility tokens are also sometimes sold during ICOs. They will, however, provide a claim on services, as opposed to security tokens. The primary distinction between a security token and a utility token is that security tokens pass the Howey Test, whereas utility tokens do not.

Stablecoins

The primary goal of stablecoins is to eliminate the volatility associated with cryptocurrencies. They mitigate volatility by tying the coin's value to real-world assets such as fiat currencies (such as the US dollar or Euro), resources (such as gasoline), and precious metals (like gold). Tether (USDT) and Libra (which may be the most popular stablecoin even though it hasn't yet launched) are two popular examples of stablecoins.

The Importance Of Altcoins

The significance of altcoins can be found in their definition: they address the limitations of bitcoin and provide alternatives with varying benefits.

Altcoins may be the perfect option for people looking for a new flavor outside of the world of Bitcoin. They have specifically addressed the issues people have with bitcoin and provided strong alternatives at comparable prices and speeds.

Another reason altcoins are significant is that they make excellent investment tools. With the success of bitcoin fresh in investors' minds, people are looking to get on the ground floor by creating and investing in new and practical cryptocurrencies—then selling them when their value rises.

The Difference Between Altcoins And Bitcoin

When it comes to cryptocurrency, Bitcoin has long been the king of the hill due to its status as the industry's founder and its first-mover advantage. A decade later, the original cryptocurrency remains the most valuable on the market, with prices reaching as high as $20,000 for a single Bitcoin at one point. It is no longer alone in the field today. As blockchain (the technology that cryptocurrency is based on) evolved, so did the number of coins available and the capabilities of the blockchains of these new coins.

These new cryptocurrencies, known as "altcoins," use the same decentralized concept as Bitcoin but go a step further by incorporating unique features. Ethereum, the second most popular cryptocurrency, pioneered the concept of "smart contracts," or code that uses blockchain technology to automatically execute agreements between two parties. This paved the way for the creation of new crypto use cases and applications.

More importantly, altcoins have improved overall functionality by processing transactions faster than bitcoin and generally scaling to meet the increasing demand for their services. As the market for Altcoins grows, it's natural to wonder if Bitcoin's lead will be short-lived or if it will be able to keep up with the next generation of cryptocurrencies.

Bitcoin was created as an alternative, decentralized digital currency that could eventually replace fiat money such as the dollar and euro. As such, it was designed for simple transactions and employs a peer-to-peer consensus mechanism to power a network that collectively verifies transactions, adding them to the "chain," which is made up of a string of transactions in batches known as blocks. As a payment mechanism, Bitcoin still lags far behind other methods such as credit cards and even other digital payment tools. Furthermore, verifying ("mining") transactions is time-consuming and costly.

Newer coins use different mechanisms to reduce mining costs and complexity. They can process far more transactions per second than bitcoin's paltry seven. Furthermore, some of these new cryptocurrencies use technology like smart contracts, which allow them to build innovative apps directly on the blockchain.

Coins such as Ripple and Dash, for example, provide a new perspective on payment transaction ability and speed. Ripple is intended to help large corporations and institutions which conduct centralized cross-border transactions. Dash claims to have transaction speeds of up to one second per transaction, as well as a focus on superior security and an easy ecosystem for individuals to manage their money.

Bitcoin, in its current form, simply cannot compete with these newer, more focused coins. Bitcoin was designed to be a catch-all currency. Its creator most likely had no idea that blockchain technology would have so many applications. Because of this imbalance, pundits and industry veterans have repeatedly claimed that Bitcoin is on its way out.

It appears that rumors of Bitcoin's demise were greatly exaggerated. Instead of fading into obscurity, it has evolved to catch up to the Altcoin market, increasing its usability. In fact, Bitcoin still has a larger user base, which comes with mainstream appeal and significant developer interest. It is now fending off newcomers by gradually adding new tools and functionality.

Rather than creating the next Bitcoin, many projects have chosen to expand on the existing Bitcoin architecture, by adding new features that make the currency more useful in various situations. RSK, for example, provides users with smart contract capabilities for Bitcoin, allowing for app development. This was once Ethereum's main selling point, which the Bitcoin is now encroaching on it, thanks to RSK's platform's expanded functionality.

Similarly, tools like the Lightning network allows the users to off-chain their Bitcoin transactions, relieving the main Bitcoin blockchain of the burden and speeding up the rate at which peripheral transactions can be verified. These solutions do not alter Bitcoin's original design but make it more competitive against younger and newer coins vying for attention. In fact, addressing these issues will only increase Bitcoin's usability and mainstream acceptance.

Although Bitcoin, in theory, could eventually do everything which Altcoins does, the reality is that it still benefits from competition. Blockchain is still a young technology that requires a thriving ecosystem to truly develop and become valuable to society. Furthermore, Bitcoin may do well to avoid feature creep and losing value. The beauty of blockchain technology is that it allows cryptocurrencies to be used for much more than just payment.

Tools such as Golem, a blockchain-powered crowdsourced supercomputer, and Fishcoin, tracks fish and seafood from the sea to millions of kitchens for ethical fishing and sustainable operations, pushing the blockchain concept in novel directions. Bitcoin is designed to be a digital currency, and it's getting better at it all the time. However, in a world where data is designed to be transactional, almost any idea will inevitably be transformed by transplanting it into a decentralized ecosystem—and Bitcoin is at the heart of it all.

Basics Of Ethereum

After Bitcoin, Ethereum is frequently referred to as the second most popular cryptocurrency. Unlike Bitcoin and most other virtual currencies, however, Ethereum is intended to be much more than a medium of exchange or a store of value. Ethereum, on the other hand, refers to itself as a decentralized computing network based on blockchain technology. Let's have a look at what that means.

How Does Ethereum Function?

Ethereum, like all cryptocurrencies, is built on a blockchain network. A blockchain is a decentralized, distributed public ledger that verifies and records all transactions.

It is distributed in the sense that everyone on the Ethereum network has an identical copy of this ledger, allowing them to see all previous transactions. It is decentralized because the network is not operated or managed by a centralized entity but by all distributed ledger holders.

Cryptography is used in blockchain transactions to secure the network and verify transactions. People use computers to "mine" or solve complex mathematical equations that confirm each transaction on the network and add new blocks to the blockchain. Participants receive cryptocurrency tokens as a reward. These tokens are known as Ether in the Ethereum system (ETH).

Ether, like Bitcoin, can be used to buy and sell goods and services. It has also experienced rapid price increase in recent years, making it a de facto speculative investment. However, what distinguishes Ethereum is that users can create applications that "run" on the blockchain in the same way a software "runs" on the computer. These programs can store and transfer personal information as well as handle complex financial transactions.

"Ethereum is unlike Bitcoin in which the network can perform computations as part of the mining process," says Ken Fromm, director of education and development at the Enterprise Ethereum Alliance. "With this fundamental computational capability, a store of value and medium of exchange is transformed into a decentralized global computing engine and openly verifiable data store."

What Is the Difference Between Ether and Ethereum?

Ether can be used in financial transactions, as an investment, or as a store of value. Ethereum is the blockchain network where Ether is stored and traded. However, as previously stated, this network provides a variety of other functions in addition to ETH.

"These can be simple fund transfers, but they can also be complex transactions that do anything from exchanging assets to taking out loans to purchasing a piece of digital art," says Boaz Avital, head of product of Anchorage's. The Ethereum network processes the transactions and stores them.

The Ethereum network can also be used for data storage and the execution of decentralized applications. People can host applications on the Ethereum blockchain rather than on a server which is owned and operated by Google or Amazon. Only one company controls the data. This gives users control over their data and allows them to freely use the app because there is no centralized authority that manages everything.

Self-executing contracts, also known as smart contracts, are among the most intriguing use cases for Ether and Ethereum. Two parties enter a contract to deliver goods or services in the future, just like any other contract. Unlike traditional contracts, no lawyers are required: The contract is coded on the Ethereum blockchain by the parties. When the conditions of the contract are met, it self-executes and delivers Ether to the appropriate party.

Ethereum vs. Bitcoin

Bitcoin's primary function is that of a virtual currency and a store of value. Although Ether can be used as a virtual currency and a store of value, the decentralized Ethereum network allows for creating and executing applications, smart contracts, and other transactions on the network. These functions are not available in Bitcoin. It is only used as a form of currency and a store of value.

Ethereum also has a faster transaction processing time. "On the Bitcoin network, new blocks are validated once every 10 minutes, while on the Ethereum network, new blocks are validated once every 12 seconds," says Gary DeWaal, chair of Katten's Financial Markets and Regulation group. And, as he points out, future developments may speed up Ethereum transactions even more.

Finally, there is no limit to the number of potential Ether tokens, whereas Bitcoin will only release 21 million coins.

Ethereum Advantages

Existing network with large size. "The benefits of Ethereum include a tried-and-true network that has been tested over years of operation and billions of dollars changing hands," says Fromm. "It has the largest ecosystem in blockchain and cryptocurrency, as well as a large and committed global community."

A wide range of functions is available. Ethereum can be used to process other types of financial transactions, execute smart contracts, and store data for third-party applications in addition to being used as a digital currency.

Constant improvement. A sizable community of Ethereum developers is always looking for new ways to improve the network and create new applications. "Because of its popularity, Ethereum is often the preferred blockchain network for new and exciting (and sometimes risky) decentralized applications," Avital (Head of product at Anchorage) says.

Avoids the use of intermediaries. The decentralized network of Ethereum promises to save the users from the involvement of third-party intermediaries such as lawyers who write and interpret contracts, banks who act as intermediaries in financial transactions, and third-party web hosting services.

Ethereum Drawbacks

Increasing transaction costs, The increasing popularity of Ethereum has resulted in higher transaction costs. Ethereum transaction fees, also known as "gas," reached a new high of $23 per transaction in February 2021, which is great if you're a miner but not so great if you're trying to use the network. Unlike Bitcoin, where the network rewards transaction verifiers, Ethereum requires those involved in the transaction to pay the fee.

Crypto inflation is a possibility. While Ethereum has an annual limit of 18 million Ether, there is no lifetime limit on the number of coins created. This could imply that, as an investment, Ethereum will behave more like dollars and will not appreciate Bitcoin, which has a strict lifetime limit on the number of coins that can be created.

Developers face a steep learning curve. As developers transition from centralized processing to decentralized networks, Ethereum can be difficult to learn.

The future is unknown. Ethereum is still evolving and improving, and the ongoing development of Ethereum 2.0

promises new functions and increased efficiency. On the other hand, this major network update creates uncertainty for apps and deals that are currently in use. According to DeWaal, “many new validators will be required for Ethereum 2.0 to function.” “The question is whether the migration will be successful. There are a lot of new elements that need to come together!”

How to Purchase Ethereum

It's a common misconception among newcomers to the Ethereum network. You don't buy Ethereum; you buy the network. Instead, you purchase Ether and use it on the Ethereum network. Because of Ethereum's popularity, it's very simple to obtain Ether:

Decide on a cryptocurrency exchange. To buy and sell various cryptocurrencies, crypto exchanges and trading platforms are used. Among the larger exchanges are Coinbase, Binance, and Kraken. Suppose you only want to buy the most popular coins, such as Ether and Bitcoin. In that case, you could also use an online brokerage like Robinhood or SoFi. Almost always, you will have to pay some sort of trading or processing fee.

Make a deposit of fiat money. To fund Ether purchases, you'll need to deposit cash, such as dollars, into your trading platform or link your bank account or debit card.

Purchase Ether. After you've funded your account, you can use the funds to buy Ether at the current Ethereum price, and use to buy other assets. Once you have the coins in your account, you can keep them, sell them, or exchange them for other cryptocurrencies in the future. Keep in mind that whenever you sell or trade cryptocurrencies, you may be subject to taxes.

Make use of a wallet. While you could store the Ether in the default digital wallet of your trading platform, this poses a security risk. If the exchange is hacked, anyone can easily steal your coins. Another option is to move coins that you do not intend to sell or trade anytime soon into another digital wallet or a cold wallet that is not connected to the internet for safety.

Should You Purchase Ether?

According to DeWaal, there are several reasons why you should consider investing in the Ethereum network. "First, it has value and use as a virtual currency; second, the Ethereum blockchain may become more appealing when it migrates to the new protocol; and third, demand for ETH may increase as more people use Ethereum distributed apps," he says.

Aside from purchasing Ether directly, you could also consider investing in companies developing applications that use the Ethereum network. If you want assistance managing your investment, you could also invest in a professional investment fund such as the Bitwise Ethereum Fund or Grayscale Ethereum Trust. However, these are currently only available to accredited investors.

Before making any significant investment in Ether or other cryptocurrencies, consult with a financial advisor about the risks involved. Given the high risk and volatility in this market, even if you believe in Ethereum's potential, make sure that you cannot afford to lose the money.

Chapter 3

The Advantages And Disadvantages Of Cryptocurrency

Any investment comes with its own pros and cons, and Bitcoin is no different. In fact, the advantages and overall risks of investing in Bitcoin are similar to cryptocurrencies. In this chapter, you will learn all about why you ought to invest in Bitcoin and a few risks associated with such investments.

Note that the risks don't invalidate an investment. They're simply things you ought to be aware of at all times. Different people have different risk tolerances. As you read this chapter, try to figure out whether such an investment is for you.

It's easy to simply highlight the immense earning potential of cryptocurrency investments. Still, a seasoned investor looks at the risks as well. Having said all that, let's look at the advantages of investing in Bitcoin and cryptocurrencies by extension.

Advantages Of Cryptocurrency

There are many advantages of investing in cryptocurrencies. While it's true that every currency has its own quirks that lend it certain advantages, the bulk of the advantages you're going to read about deals with the entire crypto space itself.

Accessibility

This advantage isn't native to cryptocurrencies or Bitcoin. However, it's worth mentioning this because these investments are far more accessible now than they were just a few years ago. In the previous chapter, you learned how Bitcoin was originally traded. Those methods still exist, but the bulk of trading occurs in regulated exchanges.

The likes of Coinbase and Binance makes it easy for people to trade cryptocurrencies and exchange them with fiat currency. There's no need for you to learn any technical wizardry or earn a computer science degree. The process of opening an account with these exchanges is straightforward, and you can open one with in few minutes.

Transparency is a side effect of accessibility. With more people trading cryptocurrencies these days, the prices of Bitcoin and other cryptos are constantly updated, and there aren't any dark markets or pools that obscure the price of a coin. For example, in the early days of Bitcoin trading, people negotiated prices on the spot.

Who knew what the exact price of a coin was? It depended on the individual's mood at the time. Electronic exchanges make price discovery a breeze and provide much-needed transparency to the markets. Another side effect of this transparency is that liquidity has increased tremendously.

Liquidity is a measure of how easily something can be traded. The more market participants there are, the more liquid an instrument is. Liquidity is important in any market since it allows for greater ranges in price. For instance, if there is only one seller in the market for Bitcoin, they can fix any price they like and dictate terms in the market. Buyers would be forced to pay whatever price that seller provides.

However, suppose there are a million sellers in the market. In that case, there are more chances of receiving a bargain or a fair price at the very least. Similarly, a multitude of buyers ensures that sellers receive fair prices for their holdings as well. Liquidity, transparency, and accessibility go hand in hand and make the markets a safer place to trade.

A major knock against crypto back in the day was how opaque the market was. Not only did no one understand how these coins worked, but they also couldn't access any market. This isn't the case now, and everyone has the opportunity to make money in cryptocurrencies.

Returns

This cannot be avoided, so let's get it out of the way. You can get rich by buying crypto. Immensely rich. A $1,000 investment in Bitcoin back in 2013 would be worth over $400,000 today. It isn't just Bitcoin that has grown exponentially. All kinds of crypto altcoins have experienced immense growth.

You read earlier about the ICO boom that occurred in the middle of the 2010s. Well, a few of those altcoins experienced immense growth after their ICOs. A good example of this is the altcoin Stratis, which raised $600,000 on the ICO and then rose 63,000%. In plain math, a $1,000 investment would be worth $630,000 today. Similarly, Spectrecoin raised $15,000 at its ICO in 2017 and then rose by 13,000%. Bitcoin's rise is tame in comparison!

One of the reasons for exponential growth is the inherent risk in these investments. Not every ICO experienced meteoric growth; therefore, it isn't as if someone could blindly invest in an ICO and then watch themselves become a millionaire.

However, considered investment in these ICOs would have resulted in massive gains after evaluating all the pros and cons. Bitcoin is the mainstream cryptocurrency and therefore doesn't provide gains of this order, but this does not mean it doesn't provide gains. After all, the price of Bitcoin rose from $15,000 in 2018 to $60,000 these days.

Cryptocurrencies are still in the earlier stages of their growth cycle. An analogy will help you understand this phenomenon better. Let's say

there's a company that has just been founded. It has a great product, and its customers love it. Its revenues grow from $100 per month to $10,000 per month within a year.

This is exponential growth, but it's hardly surprising given the small size of the company. The second year, thanks to greater investment and word of mouth, it grows from $10,000 per month in revenues to $100,000 per month. This is 10 times the growth and is still massive. However, it's still lower than the growth rate the company experienced in the first year.

This is understandable since the company's size has grown. The bigger you are, the harder it is to grow. This is simply because there isn't much room to grow. As any asset grows bigger, growth happens in smaller increments. Another way of looking at this is to compare the economies of a developed country with a developing one.

A developing country will experience faster economic growth than a developed one. Despite slower growth, the developed country remains bigger. When the developing country joins the ranks of developed countries, its growth rate declines as well. My point here is that crypto is still like a small company or a developing country.

There's still a lot of runways left for these assets to grow. Any investor who gets in right now stands to make a ton of money in the long run. It's just basic economics. The smaller size of the asset base means that investors can realize their gains in much smaller time frames compared to traditional investments. For instance, consider that the stock market grows at an average rate of 8-10% every year.

This is a pretty good growth rate, given how old the stock market is and how well established it is. In contrast, cryptocurrencies are still a new market, and most people are getting into it just now. Even this phenomenon is restricted to just Bitcoin and Ether to a certain extent.

Therefore, the profit potential for crypto is immense. Investors stand to make exponential gains in a short amount of time, thanks to the nascent stage the market finds itself in. Getting in now is the right move—waiting for the market to mature will only result in you leaving a lot of money on the table.

More ICO

How amazing would it have been for you to invest in Amazon right from the beginning when it was worth just a few pennies? Or Google, for that matter? The thing with the stock market is that great companies come by very rarely. Most businesses face a plethora of obstacles and have a rough path to profitability.

Even if they do become profitable, it takes them a long time to achieve growth that excites investors. In contrast, the cryptocurrency space is teeming with opportunities. There are multiple ICOs taking place every month, and the ethos behind most of these ICOs is sound.

Unlike a company that deals with multiple hurdles along the way, the aims of a crypto network are clear. There are no traditional customers to handle and no inherent business economic uncertainty to deal with. As a result, the path forward is clear and is fully in the developers' hands. Growth is thus easier to achieve, which explains the exponential growth of some of the ICOs you've just read about.

Combine this with the accessibility and the liquidity of the average cryptocurrency. You have a great situation where you can get on the ground floor of something great. In contrast, the shares of a startup are not liquid and are not accessible for the most part. There are private stock exchanges, but you need to have a net worth of a million dollars to even be able to participate.

There are no such restrictions in cryptocurrencies, and therefore, it's easier for an investor to make lots of money quickly.

Going With The Flow

The world is going through an immense amount of change right now. Anyone who reads the news or even browses the internet will recognize this. In a single generation, the world has gone from dreaming of the internet to considering the possibility of robots assisting with daily tasks.

All this change has brought economic instability. More people are

getting left behind thanks to technological progress, and centralized structures are no longer able to deal with the rapid rate of change. As a result, more people need security from these centralized structures running on woefully outdated modes of thinking.

Cryptocurrencies represent a new way for people to take control of their own financial destiny. They represent a way for people to secure their financial futures and safeguard the value of their money, whether they're in an unstable country or not. A citizen of America has just as many reasons to invest in crypto as a citizen of Venezuela does.

While the latter's need is more urgent, there's no doubt that fiat currencies are a dead-end in the long run. All of this makes the mainstream adoption of crypto guaranteed. Therefore, the best time to enter the space is right now.

Inflation protection

Inflation has caused many currencies to lose value over time. Almost every cryptocurrency is released with a fixed amount at the time of its launch. The amount of any coin is specified in the source code; for example, only 21 million Bitcoins are released in the world. As a result, as demand grows, its value rises, keeping pace with the market and, in the long run, preventing inflation.

Self-governing and self-managing

The governance and maintenance of any currency are critical factors in its development. Developers/miners store cryptocurrency transactions on their hardware and receive a transaction fee as a reward for doing so. Because miners are compensated for their efforts, they keep transaction records to be accurate and up to date, preserving the cryptocurrency's integrity and keeping records decentralized.

Private and secure

Cryptocurrency privacy and security have always been major concerns. The blockchain ledger is built on various mathematical puzzles that are difficult to decode. As a result, cryptocurrency transactions are more secure than standard electronic transactions. To improve security and privacy, cryptocurrency uses pseudonyms unrelated to any user, account, or stored data linked to a profile.

Currency exchanges are simple to complete

Many currencies, including the US dollar, European euro, British pound, Indian rupee, and Japanese yen, can be used to purchase cryptocurrency. One currency can be converted into another by trading in cryptocurrency across different wallets and with minimal transaction fees using different cryptocurrency wallets and exchanges.

Centralization

One significant advantage of cryptocurrency is that it is primarily decentralized. Many cryptocurrencies are controlled by the developers who use them and people who own a significant amount of coins or an organization that develops it before it is released to the market. Decentralization helps to keep the currency monopoly free and in check so that no single organization can determine the flow and value of the coin, which, in turn, keeps it stable and secure, in contrast to fiat currencies, which are controlled by the government.

Less expensive mode of transaction

One of the most common applications of cryptocurrencies is the transfer of funds across borders. The transaction fees paid by a user are reduced to a negligible or zero amount with the help of cryptocurrency. It accomplishes this by eliminating the need for third-party verification, such as VISA or PayPal. This eliminates the need to pay any additional transaction fees.

A quick method of transferring funds

Cryptocurrencies have always been regarded as the best option for transactional purposes. Cryptocurrency transactions, whether international or domestic, are lightning-fast. This is because the verification takes very little time to complete. After all, there are very few barriers to overcome.

Disadvantages Of Cryptocurrency

While all these advantages make investing in crypto very attractive, you would be wise to consider the risks as well. Here are some of the most pertinent risks that crypto investors face.

Regulation

Right now, Bitcoin activity is unregulated. While this is a great thing if you want to protect yourself from centralized power structures, it isn't good news if you're looking at practically using your coins. A big reason for this is the nascent stage in which all cryptocurrencies find themselves.

The stock market is a heavily regulated place, which means that every investor, whether small or institutional, plays by the same rules. Just because someone has greater resources doesn't guarantee them more profits. Bitcoin doesn't quite have this protection.

While there isn't any insider information to trade, larger investors with more resources are a discouraging factor to most people. As an example, there have been reports of the Chinese government setting up massive Bitcoin mining farms that allow them to own more of the cryptocurrency.

No single investor can ever compete with a government that controls everything its people do. This makes investing in Bitcoin quite risky. The other fact is that no one knows who or what Satoshi Nakamoto is. While trading mechanisms are transparent, there is a complete lack of transparency about who this actor is, and their motivations are.

Who's to say Nakamoto won't sell their holdings due to a change of heart? Given the size of their holdings, it's fair to say that the price of Bitcoin will crash downward quickly. Will it rise back up once it crashes? It's tough to say. On the one hand, it's an opportunity to buy at lower prices; however, the founder of the coin selling out would be a pretty telling action.

Therefore, a few transparency risks inherent to Bitcoin aren't possible to overcome right now. Regulation is increasing, but investors are still left unprotected in case of fraud.

Volatile Nature

One of the major turn-offs for investors is that cryptocurrencies are extremely volatile. Volatility is a double-edged sword. On one hand, volatility ensures prices rise astronomically. On the other hand, it also means that prices will bounce around a lot. In some cases, they might bounce to extreme levels that make no sense.

As a result, you might end up entering the market at the wrong time

and lose money. Bitcoin is quite volatile, experiencing price rises and drops of over 10% per day. Given that its price is currently north of $50,000, that's a significant degree of fluctuation to deal with.

This extreme volatility is a result of the early-stage nature of Bitcoin and all cryptocurrencies. With great rewards come great risks as well. Therefore, before investing money into cryptocurrencies, you should always make sure you've covered your risks fully.

Sinking everything you have into cryptos or Bitcoin is the wrong move to make. You're just as likely to end up making a ton of money as you are to lose everything you have. Always risk only what you can afford to lose, and don't bank on cryptos being able to pay your rent or other daily expenses.

Network Stall

A significant risk with any crypto investment is that of network stall. This refers to currency's founders cannot execute their aims or if the market rejects the utility of their aims. For instance, if a network is founded on the premise of executing smart contracts, the founders need to ensure their contracts are robust and provide people with the right tokens (coins) in exchange for execution.

If anything goes wrong with the network, investors will lose a lot of money in those tokens. In essence, this risk is the same as what any investor in a business face. However, given the lack of regulation in the crypto space, investors won't have any recourse, unlike the stock market.

In the stock market, if a company fails to meet its obligations and its stock price starts sliding, investors are still able to sell stock in the market and find willing buyers. These buyers are usually large institutions who seek to buy failing companies on the cheap and turn them around.

No such recourse exists in the crypto space since institutions aren't as active. This being the case, a crypto investor wages a lonely battle that requires them to always be vigilant.

Fraud And Scams

There's no denying that the lack of regulation in the cryptocurrency space has led to several fraudulent schemes being concocted. The problem with these schemes is that they always sound great and even reasonable, but it's hard to tell what's true and what isn't. While the outright ridiculous tales of crypto fraud gain a lot of attention, the truth is that these incidents are in the minority.

What's far more prevalent is the scheme that sounds great but just isn't quite right. These schemes don't appear to be fraudulent at first glance but often turn out to be fraud at the later point due to faulty assumptions on the investor's part. Even robust currencies can come under this category because of the sheer impracticality of their aims.

I've already highlighted Monero (XOM) as a currency that seeks to improve on Bitcoin's aims. The problem with XOM is that its aims almost certainly ensures it will never be adopted by the mainstream. Imagine a bank being willing to authorize a transaction where they can't see the amount being transferred or who it's being transferred to!

The idea that any government looking to stamp out corruption and financial crime will ever let something like this run unchecked is hard to fathom. Hence, middle-of-the-road cryptos like Bitcoin and Ether do very well. They have just enough privacy-related features to make them decentralized. Still, they retain a kernel of centralization that ensures their mainstream adoption is realistic.

The reason you'll see unrealistic assumptions such as the ones made by the XOM team in the crypto space is that many founders are idealists who lack real-world financial experience. This sounds harsh, but it's true. Being extremely adept at writing code does not make someone a financial genius.

An ideal coin will have someone with technical chops and a team that takes financial realities into account. There's no doubt that centralization is hampering ordinary people, but this doesn't mean extreme deregulation is a solution. The stock market offers ample examples of what happens when deregulation is written into corporate rules. Fraud and losses are almost guaranteed.

Many people in the crypto space fail to recognize this and hold on to a utopian dream of everyone treating each other fairly and of bad actors simply vanishing into thin air. It's safe to say such people don't live in the real world. Any coin backed by such idealistic aims is unlikely to succeed. Your task is to watch out for such projects and steer clear of them.

It can be used for illegal transactions

Because cryptocurrency transactions have high privacy and security, it is difficult for the government to track down any user by their wallet address or keep track of their data. Bitcoin has previously been used to exchange money in a variety of illegal transactions, such as purchasing drugs on the dark web. Some people use cryptocurrency to convert their illegally obtained money through a trustworthy intermediary, thereby concealing its source.

Financial losses can result from data loss

The developers desire virtually untraceable source code, robust hacking defenses, and impenetrable authentication protocols. This would make storing money in cryptocurrencies safer than in physical cash or bank vaults. However, if a user loses the private key to their wallet, there is no way to recover it. The wallet, as well as the coins inside it, will be kept secure. The user will suffer a financial loss as a result of this.

Decentralized but still operated by some organizations – Cryptocurrencies are well-known for their decentralized nature. However, the flow and amount of some currencies in the market are still under the control of their creators and certain organizations. These holders can manipulate the coin's price to cause large price swings. Even highly traded coins, such as Bitcoin, which more than doubled in value in 2017, are vulnerable to these manipulations.

Some cryptocurrencies cannot be traded in other fiat currencies

Some cryptocurrencies can only be traded in one or a few fiat currencies. This forces the user to first convert these currencies into one of the major currencies, such as Bitcoin or Ethereum, and then to their desired currency via other exchanges. This only applies to a few cryptocurrencies. By doing so, extra transaction fees are added to the process, costing extra money.

Negative Mining's environmental impact

Mining cryptocurrencies requires a lot of computational power and electricity input, making it a very energy-intensive process. Bitcoin is the main culprit in this.

Mining Bitcoin necessitates powerful computers and a lot of energy. It is not possible to do so on standard computers. Major Bitcoin miners are in countries that use coal to generate electricity, such as China. This has significantly increased China's carbon footprint.

Susceptible to hacks

While cryptocurrencies are extremely secure, exchanges are not. Most exchanges save users' wallet data to properly operate their user ID. Hackers can steal this information and gain access to many accounts.

These hackers can easily transfer funds from those accounts once they have gained access. Some exchanges, such as Bitfinex and Mt Gox, have been hacked in recent years. Bitcoin worth thousands and millions of dollars have been stolen. Most exchanges are now highly secure, but there is always the possibility of another hack.

There is no refund or cancellation policy

If there is a dispute between the parties involved, or if funds are sent to the wrong wallet address by mistake, the coin cannot be retrieved by the sender. Many people can use this to defraud others of their money. As there are no refunds, one can easily be created for a transaction for which they never received the product or services.

Chapter 4

How To Start Cryptocurrency

What To Know Before Starting Your Business?

Starting a business is never an easy task, especially when it involves cryptocurrency. Cryptocurrencies continue to occupy a new market space that is influenced by financial and regulatory risk. Stories of fame, fortune, and get-rich-quick IPOs, on the other hand, entice many an entrepreneur to enter the cryptocurrency sector — and all too often for the wrong reasons.

Cryptocurrency has the potential to democratize finance by establishing a true peer-to-peer medium of exchange that is not subject to government seizure. Bitcoin is a liberating technology that has the potential to transform the world. Suppose you agree with that message and truly believe it, then I believe you have the right motivation to begin a career in cryptocurrency.

I've spent the better part of a decade working with cryptocurrencies as an alternative investment professional. In this article, I've discussed a few key things you should be aware of if you want to get started as a cryptocurrency entrepreneur and contribute to the advancement of alternative finance.

Don't Underestimate Technical Skills

If you want to start a crypto company, you or your co-founders must be technically savvy. Whether you want to be a developer, broker, or marketer, every crypto company requires technical skills to fully understand the underlying blockchain systems.

- Understand how distributed ledgers work (without the jargon!).
- Be familiar with smart contracts and the Ethereum protocol.
- Be familiar with technical and fundamental financial market analysis.
- Have a working knowledge of Solidity, JavaScript, and TypeScript.

Even if you don't want to start a business as a developer, you'll almost certainly have to make technical claims as a bitcoin entrepreneur or crypto founder. You'll need some technical know-how to back it up.

Understand How To Attract Venture Capitalists

It goes without saying that cryptocurrencies are volatile and pose a significant financial risk to you and your early investors. Regardless of the risk, you must be able to approach and persuade venture capital (VC) firms to invest in your project.

Is equity financing not an option for you? If you're specifically interested in creating your own token, you can always go the initial coin offering (ICO) route. This, however, comes with its own set of serious financial risks and legal liabilities that you must be aware of.

The bottom line is that you will require interpersonal skills. Practicing public speaking and presentation skills will greatly help you in your crypto entrepreneurship, as several blockchain-based venture capitalists are willing to listen to your project.

Be Flexible In Your Approaches To Innovation And Adaptation

There are no skills which is more important than the ability to innovate and adapt in such a fast-paced industry. When I first got involved in the crypto space, Ethereum (ETH), the No. 2 most valuable blockchain technology by market cap, had not been conceived. It now accounts for a sizable portion of my crypto business.

When the market shifts, which can happen seemingly overnight, you must adapt quickly. If you don't, your competitors will, and you'll lose market share. You can expect to put in endless days of work developing the technical side of your operation to keep up with the changing environment during particularly chaotic periods in the industry.

Similarly, blockchain and innovation are nearly synonymous terms. Suppose you want to stay afloat in this highly competitive market, you must be willing to experiment with new revenue streams and offer products and incentives that your competitors will not.

Increase Your Risk Tolerance

Blockchain is a high-risk, high-reward industry that should not be taken lightly. Blockchain, as a nascent technology, is subject to speculation and vulnerable to swings in investor sentiment. If the price of Bitcoin falls as it did in January 2018 (-65%) or April 2013 (-71%), your business may take years to recover.

Diversifying your offerings as a crypto entrepreneur is one of the best things you can do to avoid becoming overly exposed to crypto. For example, I work in the alternative assets space, including precious metals and other asset classes unrelated to Bitcoin. Risk-averse entrepreneurs would be wise to diversify their revenue streams, to face market downturns.

Keep in mind that we're all just getting started.

According to a recent research study, the global blockchain market will be worth more than $60 billion by 2025, representing a 69.4 percent compound annual growth rate. Blockchain and cryptocurrencies are still in their infancy, according to most accounts. As a result, the market can and will change over time.

What is true for cryptocurrency entrepreneurs today may not be true tomorrow. The advice I've given above may not even be applicable in a few years, let alone months. This, I believe, speaks to an important aspect of crypto entrepreneurship that every aspiring business owner in the space should be aware of: You must be willing to accept risk and always be willing to adapt to a changing market if you want to succeed in this industry.

Understand The Difference Between A Coin And Token

To begin, it is critical to understand the distinction between coins and tokens. While both are cryptocurrencies, a coin—Bitcoin, Litecoin, Dogecoin—operates on its own blockchain. A token lives on top of an existing blockchain infrastructure, such as Ethereum. In the simplest terms, A blockchain is a record of transactions made and secured by a network. While coins have their own independent transaction ledgers, tokens rely on the technology of the underlying network to verify and secure transactions and ownership. Coins are used to transfer wealth in general. In contrast, tokens can represent a "contract" for almost anything, from physical objects to event tickets to loyalty points.

Tokens are frequently distributed through a crowd sale known as an initial coin offering (ICO) in exchange for existing coins, which are then used to fund projects such as gaming platforms or digital wallets. You can still buy publicly available tokens after an ICO has ended, just like you can buy coins by using the underlying currency.

Anyone can create a token and hold a crowd sale, but ICOs can be shady if creators take investors' money and flee. Some ICOs are considered securities and are thus regulated by the Securities and Exchange Commission, which warns investors to conduct their own research before purchasing tokens launched in an ICO. The SEC is cracking down harder on fraudulent ICOs.

CoinMarketCap lists over 5,200 coins and tokens available on public exchanges at the time of writing. However, not all tokens made it to exchanges — Etherscan, a provider of Ethereum analytics, has over 403,000 contracts in its archive. In April 2018, the figure was only 71,000.

The idea behind cryptocurrency is that the underlying code is open to all—but that doesn't mean it's simple to grasp. Here are the steps to making your own coins and tokens.

The Advantage Of Having Your Own Cryptocurrency

In some cases, it is self-evident: if your project or startup requires its own blockchain, you must create your own digital currency to incentivize the nodes contributing their processing power. Another word on blockchains: many credible business analysts predict a bright future and a growing list of markets and industries where blockchain technology will significantly disrupt the status quo and generously reward early adopters. The good news is that blockchain technology has not yet truly arrived in many fields, so it is not too late to join the ranks of pioneers.

Another important aspect is that when you decide to launch a cryptocurrency, you gain access to a slew of powerful marketing tools and consumer benefits that will help you stand out from the crowd. Here is a list of the most important benefits:

Eliminating fraud risks — cryptocurrency cannot be counterfeited, and no party can reverse previous transactions. The fact that it cannot be counterfeited and the increased security gives users total control over the information that they want to share while making a transaction.

Lowering operating costs — cryptocurrency is free of exchange and interest rates, as well as transaction fees.

Offering instant transactions — state holidays, business hours, or the parties' geographic location have no bearing on cryptocurrency.

Providing an immediate pool of potential customers — you can now conduct business with those who do not have access to traditional exchange resources. There will be no more trade restrictions in any markets. Providing funds security — because cryptocurrency is a decentralized system, there is no Big Brother figure like banks or government institutions that can seize or freeze your assets.

The Steps To Creating Your Own Cryptocurrency

Now that you understand how your own cryptocurrency can help your business, let's go over the main steps you'll need to create a blockchain.

Step 1: Understand Your Use-Case Scenario

Do you have a vested interest in smart contracts, data authentication, verification, or smart asset management? Define your objectives clearly from the start.

Step 2: Pick A Consensus Mechanism

To ensure the smooth operation of your blockchain, the participating nodes must agree on which transactions should be considered legitimate and added to the block. Consensus mechanisms are protocols that accomplish this There are numerous options available to find the best fit for your business objectives.

Step 3: Pick Your Desired Blockchain Platform

The consensus mechanism you choose will influence your choice of a blockchain platform. Here is a list of the most popular blockchain platforms to give you a better idea of what is available:

Ethereum (market share—82.70%)
Waves (WAVES)
NEMNxt (NXT)
BlockStarter
EOS
BitShares 2.0
CoinList
Hyperledger Fabric
IBM blockchain
MultiChain
HydraChain
BigChain
DBOpenchain
Chain Core
QuorumIOTA
KICKICO

Step 4: Design The Nodes

If you think of a blockchain as a wall, nodes are the bricks that make it up. A node is an Internet-connected device that helps a blockchain by performing various tasks such as data storage, transaction verification, and processing. Nodes are essential to the efficiency, support, and security of blockchains.

You have several options when it comes to the nodes you can use:

- In terms of permissions, will they be private, public, or hybrid?
- Will they be in the cloud, on-premises, or both?
- Choose and obtain the necessary hardware details, such as processors, memory, disk size, and so on.
- Choose a base operating system (most common choices would be Ubuntu, Windows, Red Hat, Debian, CentOS, or Fedora)

Step 5: Establish A Fine Internal Architecture For Your Blockchain

As some of the parameters cannot be changed once the blockchain platform is up and running, proceed with caution. It's a good idea to take your time and thoroughly consider the following:

- Permission (define who can access the data, perform transactions and validate them, i.e., create new blocks)
- Address Formats (decide what your blockchain addresses will look like)

- Key Formats (decide on the format of the keys that will be generating the signatures for the transactions)

- Asset Issuance (establish the rules for creating and listing all asset units)

- Re-issuance of assets (establish the rules for creating more units of the open assets)

- Key Management (develop a system to store and protect the private keys granting the blockchain access)

- MultiSignatures (define the number of keys your blockchain will require to validate a transaction)

- Atomic swapping (plan for the smart contracts enabling the exchange of different cryptocurrencies without a trusted third party)

- Parameters (estimate maximum block size, rewards for block mining, transaction limits, etc.)

- Native resources (define the rules of a native currency issued in a blockchain)

- Signatures that are blocked (define how the blockchain participants creating blocks will be required to sign them)

- Handshakes (establish the rules of how the nodes will identify themselves when connecting to each other)

Step 6: Manage The API's

Check to see if the blockchain platform of your choice offers pre-built APIs, as not all of them do. Even if your platform does not include these, do not worry: there are many trustworthy blockchain API providers out there.

Step 7: Carefully Design The Interface

Communication is essential, and a well-thought-out interface ensures that your blockchain and its participants communicate smoothly.

Here are some things to think about at this point:

- Servers for web, email, and FTP

- External data sources

- Front-end programming languages (for example, HTML5, CSS, PHP, C#, Java, Javascript, Python, Ruby)

Step 8: Make Your Crypto Legal

Slowly but steadily, the law is catching up with cryptocurrencies. You should prepare for any surprises by researching the trends surrounding cryptocurrency regulations and their direction. Overachievers should take an extra step to grow and improve their Blockchain.

A Note On Bitcoin Forking

As you can see, creating a blockchain requires a significant amount of time, resources, and specialized knowledge. So, what can you do if you don't have all the resources above but still want to create your own cryptocurrency? Then we'll talk about Bitcoin forks.

How Do You Make a Bitcoin Fork?

It's time for another fundamental definition to ensure that we're all speaking the same language.

What exactly is cryptocurrency forking?

A blockchain fork is a software update in layman's terms. All blockchain participants (also known as full nodes) use the same software. They all must use the same version of the software to access the shared ledger, verify transactions, and ensure network security. As a result, every time you want to change your blockchain parameters or add new features, you must create a fork.

What's the distinction between hard and soft forks?

There are two types of forks: hard and soft.

Hard forks necessitate updating 90 to 95 percent of the nodes' software; the system will no longer accept nodes running an un-updated version.

Soft forks are easier to use. Only a majority of the nodes are required to update the software. Those running an earlier version can continue to function.

What exactly are Bitcoin forks?

Bitcoin forks are now defined as changes to the Bitcoin network protocol. Because the Bitcoin code is open-source, it is a low-lift exercise for those who want to create their own cryptocurrency and build on the Existing cryptocurrency codes and improve upon them.

How do you make a Bitcoin fork?

Option 1: Make use of a fork coin generator.

If you don't have any programming skills, services like ForkGen may be ideal for you. ForkGen is an automated fork coin generator that allows anyone to create a one-of-a-kind Bitcoin offshoot by modifying some parameters and rules.

Option 2: Do It Yourself

If you want to create a Bitcoin fork the hard way and aren't afraid to get your hands dirty, follow these steps:

- Go to Github, find Bitcoin code, download it, and compile it on your computer.
- Then the programming begins: you must reconfigure the Bitcoin code and implement your customization.
- Return the code to Github as open source.
- Make a website and documentation available (normally a white paper).
- Success stories from Bitcoin forks

Bitcoin forks are worth investigating if you want to launch your own cryptocurrency while capitalizing on the social and financial capital associated with the Bitcoin brand. Here are some examples of successful Bitcoin forks:

- CashBitcoin
- Goldman
- LitecoinBitcoin

Chapter 5

How To Buy Cryptocurrency

What To Know Before Buying

Over the last few months, the process of buying and selling cryptocurrency has been greatly simplified. There are five critical factors that you must consider.

Location

You must check your country's regulations to learn how and where you can buy cryptocurrency.

Method Of Payment

Credit cards, bank transfers, or even cash are the most common and accepted payment methods for purchasing cryptocurrency. Different websites accept different payment methods. You must select a website that accepts the payment method you wish to use.

Types Of Cryptocurrency

Not all cryptocurrencies can be purchased on every website. You must first locate a website that sells the cryptocurrency that you wish to purchase.

Fees

Each website has its own set of fees. Some are inexpensive, while others are not. Before creating an account on any website, make sure you understand the fees. You don't want to waste time verifying yourself only to discover that the fees are prohibitively expensive!

Your Affordability

You should never invest more than you can afford, as with any other investment. I recommend first consulting with a financial advisor.

Choosing A Wallet

A cryptocurrency wallet is where you keep your cryptocurrencies once you've purchased them. A cryptocurrency wallet can be compared to a bank account. You will store your cryptocurrencies in your crypto wallet in the same way that you store traditional currencies (USD, JPY, EUR, etc.) in your bank account.

There are numerous simple and secure options to choose from. You must select a highly secure wallet because if your cryptocurrency is stolen from your wallet, you will never be able to recover it.

Wallets are classified into three types:

- Online wallets are the easiest to set up (but also the least secure);
- Software wallets are apps that you download (safer than online wallets).
- Hardware wallets are portable devices that connect to your computer via USB (the safest option).

The wallet you require will be determined by the cryptocurrency you wish to purchase. If you want to buy Bitcoin, for example, you'll need a Bitcoin wallet. If you want to buy Litecoin, you'll need a wallet that can hold it.

Fortunately, there are numerous good wallets available that can store multiple cryptocurrencies. Exodus, for example, is a software wallet that can hold Bitcoin, Ethereum, Litecoin, Dash, EOS, and other cryptocurrencies.

I recommend the Ledger Nano S. It is extremely safe (because it connects to the internet only when plugged into your computer) and costs around $70. It holds NEO, Bitcoin, Ethereum, Bitcoin Cash, and many other cryptocurrencies!

Wallets are discussed to a greater extent in the next chapter.

Where Should You Buy Cryptocurrency?

To learn how to buy cryptocurrency, you must first understand where you can buy cryptocurrency. There were only a few places to buy and sell cryptocurrencies a few years ago. But now there are a lot more! Let's look at where and how you can get your crypto.

How Do You Buy Cryptocurrency At Exchanges?

Cryptocurrency exchanges are websites where you can exchange your local currency for cryptocurrency. The most common method of purchasing cryptocurrency is through exchanges. I recommend that newcomers use these exchange websites because they are simple to use and quick to set up.

Payment Method: Most exchanges accept payment via bank transfer or credit card, with few also accepting PayPal.

There are numerous exchanges from which to choose. Each of them has varying levels of security and accepts a variety of payment methods. Most exchanges will require you to complete the steps outlined below before you can begin purchasing cryptocurrency.

- The first step is to provide your information (full name, email address, cell phone number, etc.).
- The next step is to verify your identity, which usually requires a passport or a driver's license.
- Following that, you will be prompted to enable 2-Step Verification. This helps to keep your account safe and hackers at bay!

It's worth noting that almost all exchanges charge a transaction fee of 1-4 percent.

After you've created an account, you can begin purchasing cryptocurrency on the exchange. Most exchanges function similarly to stock trading platforms in that you can buy and hold your cryptocurrency or exchange it for another cryptocurrency.

When choosing a cryptocurrency exchange, you must exercise caution. You should base your decision on the following considerations:

Examine the exchange's security. This is the most important factor to consider, as investors had lost a lot of money in the past when exchanges were hacked. The lower the transaction fees, the better.

The variety of payment options: do they accept bank transfers, credit cards, PayPal, and so on?

The time it takes to activate your account — if you're in a hurry, you might not want to wait too long to be verified!

What cryptocurrency options they provide. For example, if you want to purchase the NEO coin, you must ensure that the NEO coin exchange which you select sells as well.

Examine the exchange's user reviews! This will give you an idea of how user-friendly the exchange website is.

The Best Exchange Brokers

Broker exchanges are similar to currency exchange shops found in airports. However, rather than exchanging different local currencies (for example, YEN to USD), you can exchange your local currency for cryptocurrencies. It's the simplest way to buy cryptocurrency, and you can even do it from your phone!

Coinbase

Coinbase is the largest and most well-known cryptocurrency brokerage exchange. It is safe and secure, which accepts bank transfers, credit/debit cards, and PayPal payments. Their fees vary depending on the payment method you select, so double-check before you pay! Card fees are the most expensive, at 3.75 percent. If you live in Asia, Africa, or South America, you will most likely be unable to use Coinbase. It is primarily sold in Europe, Australia, and North America. Coinbase allows you to purchase Bitcoin, Ether, Litecoin, and Bitcoin Cash.

Coinmama

Coinmama's main advantage is that it allows customers from almost any country to buy and sell cryptocurrency. It is simple to use and well-known — it was founded in 2013! It does accept credit cards as a payment method, but there is a 6% fee for doing so. This is significantly higher than Coinbase's 3.75 percent fee! Coinmama, in addition to being available in almost every country, has another significant advantage: it accepts every local currency!

On Coinmama, you can only buy Bitcoin and Ether.

CEX.io

CEX.io, like Coinbase and Coinmama, is simple to use and extremely secure! The best part about CEX.io is that it also functions as a trading exchange. This means that once you've purchased your cryptocurrency, you can trade it for other cryptocurrencies by buying and selling it to other users on CEX.io.

Another feature that distinguishes CEX.io is that it accepts USD, EUR, GBP, and RUB! On CEX.io, debit/credit card fees are 3.5 percent + $0.20. You can also send money through a bank transfer. CEX.io allows you to purchase Bitcoin, Ethereum, and DASH.

Bitstamp

Bitstamp is one of the oldest cryptocurrency exchanges, which was founded in 2011. Bitstamp allows you to purchase a wider range of cryptocurrencies than Coinbase. It accepts bank transfers as well as credit and debit cards in EUR and USD.

However, I would not recommend this website for complete beginners because it is not as user-friendly as Coinbase and Coinmama. It also charges exorbitant fees for making small purchases with your credit card.

When you use your debit card on Bitstamp, the fees are 2% if you deposit more than $1,000 and $10 if you deposit less than $1,000. I would not recommend using Bitstamp to spend only $5!

Bitstamp allows you to purchase Bitcoin, Ether, Litecoin, Bitcoin Cash, and Ripple.

Kriptomat

Finally, Kriptomat is an exchange that has only recently opened its doors but has great potential. Kriptomat is the newest brokerage on our list, having opened its doors in 2018.

Their main operation hub is in Slovenia, but the company is based in Estonia. Estonia is one of the few countries with the necessary legal framework. It has already begun to regulate digital currencies by introducing licenses for Crypto Exchanges and Crypto Wallets.

Kriptomat aims to provide the most secure fiat gateway into the world of cryptocurrencies and has already received operating licenses from the Financial Intelligence Unit, the Estonian Anti Money Laundering (AML) authority.

Because of the strict regulation, users must go through a rigorous process of identity verification. However, on Kriptomat, this process is very simple. It takes only 5 minutes of your time to upload all the necessary documents.

In some countries, you can even do it from your phone!

The buy/sell fees on Kriptomat are 1.45 percent.

Keep in mind that apart from exchanges, you may also buy bitcoin from

- ATM machines
- Peer To Peer Networks

Both of which has their pros and cons. But choosing the perfect one for you will depend on your location and availability.

Each year, more ways to buy cryptocurrency becomes available — it is much easier to buy cryptocurrency now than it was previously. In fact, if you follow our instructions, most of the methods I've shown you in this guide are very simple.

But which path is the best...? Is it through broker exchanges such as Coinbase? Through peer-to-peer websites such as LocalBitcoins? Or is it as simple as finding a Bitcoin ATM near you? The truth is that it all depends on your location and what you want/need. If you don't live near a Bitcoin ATM, this is probably not the best place for you to buy cryptocurrency. You don't want to use your ID? Then you might want to consider using a peer-to-peer website, such as LocalBitcoins or LocalEthereum.

If using your ID isn't an issue for you, and you want to buy cryptocurrency using an app on your phone, Coinbase may be the best place for you to do so.

It is entirely up to you. I hope you found this guide useful in determining which path is best for you and that you find our instructions simple and easy to follow.

Chapter 6

How Do You Start Crypto Trading?

Understanding Trading

Trade is a fundamental economic concept that involves the buying and selling of goods and services, with a buyer compensating a seller or the exchange of goods or services between parties. Within an economy, trade can occur between producers and consumers. International trade allows countries to open markets for goods and services that would otherwise be closed. This is why an American consumer can choose between a Japanese, German, or American automobile. As a result of international trade, the market has more competition and more competitive prices, resulting in a cheaper product.

How Does Trading Work?

In its broadest sense, trade refers to transactions ranging in complexity from the exchange of baseball cards between collectors to multinational policies establishing protocols for international imports and exports. Regardless of the transaction's complexity, trading is facilitated by three primary types of exchanges.

Global trade allows consumers and countries to be exposed to goods and services that are not available in their own countries. Food, clothing, spare parts, oil, jewelry, wine, stocks, currencies, and water are all available on the international market. Tourism, banking, consulting, and transportation are all traded services. An export is a product that is sold to the global market, and an import is a product purchased from the global market. In the balance of payments, imports and exports are accounted for in a country's current account.

International trade increases efficiency and allows countries to participate in a global economy, which encourages foreign direct investment (FDI), which is the amount of money that individuals invest in foreign companies and other assets. In theory, economies can thus grow more efficiently and become more competitive economic participants. For the receiving government, FDI is a means of bringing in foreign currency and expertise. These increase employment and, in theory, lead to an increase in the gross domestic product. For the investor, FDI provides opportunities for company expansion and growth, which results in higher revenues.

A trade deficit occurs when a country spends more on aggregate imports from other countries than it earns from aggregate exports. A trade deficit denotes the flow of domestic currency to foreign markets. This is also referred to as a negative trade balance (NTB.

Understanding The Concept Of Investing

Investing is the allocation of resources (such as capital) with the expectation of profit. This can include using money to fund and kickstart a business or purchasing land to resell it at a higher price later. This usually entails investing in financial instruments in the financial markets to later sell them at a higher price.

The expectation of a return is central to the investment concept (also known as ROI). Investing, as opposed to trading, typically takes a longer-term approach to wealth accumulation. An investor's goal is to accumulate wealth over a long period (years or even decades). There are numerous ways to accomplish this, but most investors will rely on fundamental factors to identify potentially profitable investment opportunities.

Because of the long-term nature of their approach, investors are usually unconcerned about short-term price fluctuations. As a result, they will typically remain relatively passive, not overly concerned with short-term losses.

The Difference Between Trading And Investing

Profits in the financial markets are sought after by both traders and investors. Their approaches to achieving this goal, on the other hand, are quite different.

In general, investors seek to generate a return over a longer period, such as years or even decades. Because investors have a longer time horizon, their expected returns on each investment are typically higher.

Traders, on the other hand, attempt to profit from market volatility. They enter and exit positions more frequently. They may seek lower returns on each trade (because they frequently enter multiple trades).

Which is superior? Which one is better suited for you? It is entirely up to you. You can begin by educating yourself about markets, and then you can learn by executing. You'll be able to determine which one best suit your financial goals, personality, and trading profile over time.

What You Should Be Aware Of

The most important thing to remember before you begin trading is that you may lose your entire investment.

The cryptocurrency markets are extremely volatile, and while some people have made a lot of money, many others have lost a lot of money as well. You should never trade with money you cannot afford to lose.

Your success as a trader will be determined by how you handle losses. Here's some sound advice: never try to recoup your losses by investing larger sums. This is emotional investing, and it frequently leads to large losses.

FUD

FUD is a popular term in both real-world and crypto trading. This abbreviation stands for Fear, Uncertainty, and misinformation.

FUD is when people or organizations try to scare people away from investing in an asset by telling them they will lose all their money if they invest (or something similar). They usually say things like "It's a scam" or "It's about to crash."

Before jumping to any conclusions, you should always conduct your own research. Check Google to see if the information you're hearing is correct.

Persuasion

Persuasion is the process of convincing someone to do something.

One final thing to keep in mind before you begin trading is that you should never be swayed by the opinions of others. Remember that you are not the only one interested in profiting from cryptocurrency trading.

Be cautious of the YouTubers you follow and listen to. They are frequently compensated by cryptocurrency projects to promote their coin. This could raise the price in the short term but decrease in the long run. As a result, always conduct your own research first.

If you've read this far, you should have a good understanding of what cryptocurrency trading is, know the difference between short-term and long-term trading, and the things you should avoid.

What do you think? It's finally time to learn how to trade cryptocurrency!

With over 1500 crypto coins, the crypto market cap is now around $300 billion. Unsurprisingly, an increasing number of people worldwide are deciding to enter this industry and begin investing in cryptocurrency. However, trading cryptocurrency can be confusing because the process differs from trading on fiat exchanges,traditional investors and traders are used to.

Today, we'll go over the four essential steps that will get you started with cryptocurrency trading.

Step 1: Setting Up The Platform

We need to get you some coins first because you want to trade cryptocurrencies! The simplest way to do this is with Bitcoin, which is accepted by almost every exchange.

If you prefer to buy Ethereum, you can still follow the steps outlined below. However, whenever Bitcoin is mentioned, it should be replaced with Ethereum.

If you already have Bitcoin, you can proceed to the next section of the guide!

Using your debit or credit card with Coinbase is the quickest way to purchase Bitcoin. Coinbase is an exchange broker which sell Bitcoin at a very reasonable price. Although there is a 4% fee for using your card, it is well worth it because you get your coins right away

Opening Up The Account

Keep in mind that it is recommended that you have a strong username and password to open your account. Once you have opened your account, the next step is to verify it.

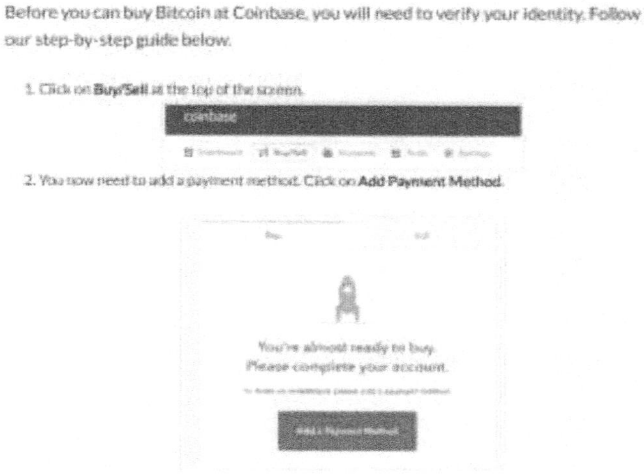

3. You will then be asked if you want to add a bank account or a debit/credit card. In this example, we will add a debit/credit card as it is the quickest way to deposit.

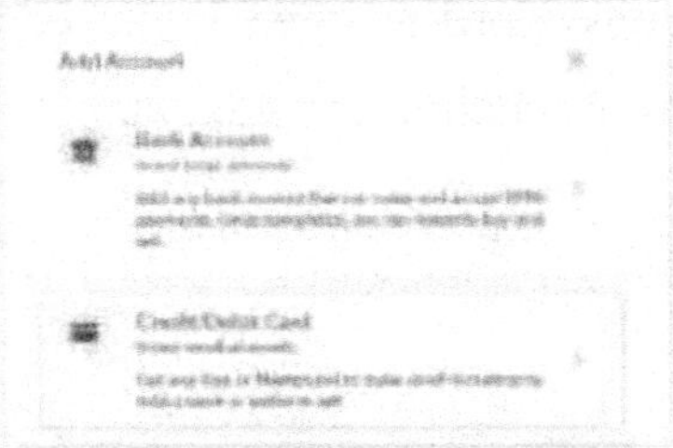

4. As Coinbase takes security very seriously, you will now need to verify your identity. Click on **Upload ID**.
5. You can upload either a passport, driver's license or a government-issued ID card. Click on the one that you wish to upload.

6. Next, you will be asked how you want to upload your ID. You can choose a webcam, mobile camera or a file upload.

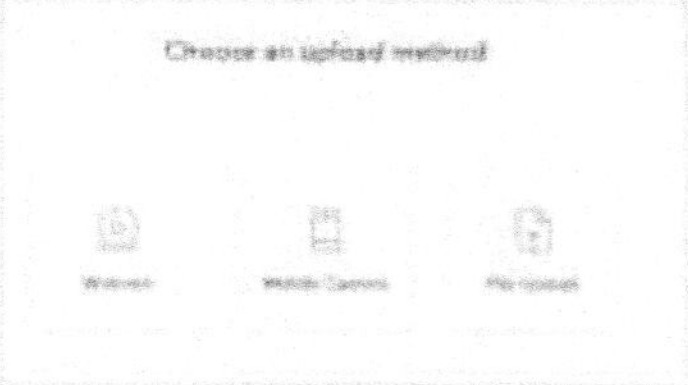

7. Once you have uploaded your ID, you will get the below confirmation.

If the picture quality if not clear enough, you will be requested to upload it again. **Add your payment method in Coinbase**

1. You will now need to enter the billing address for your debit/credit card

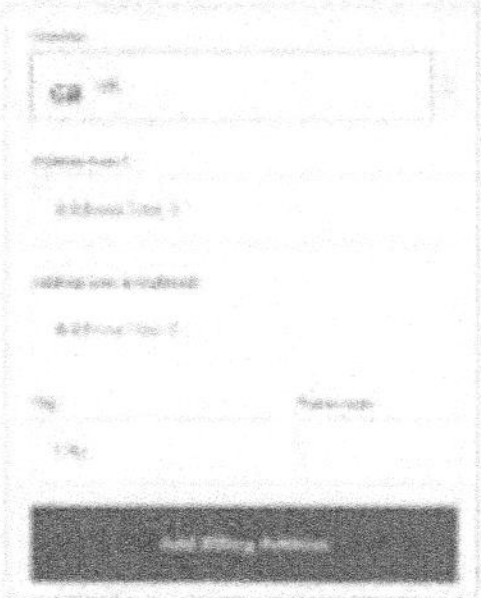

2. Now you will need to enter your debit/credit card details. Don't worry – Coinbase never gets to see your card details as the numbers are encrypted.

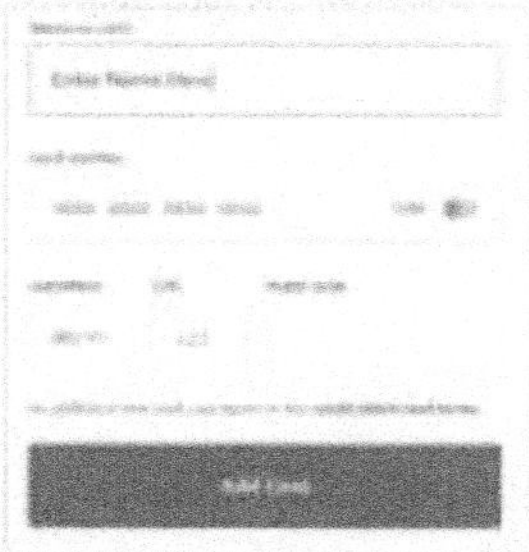

3. Congratulations, you have now verified your identity and added a payment method! Now, let's go and buy some Bitcoin.

Buying Bitcoin Through Coinbase

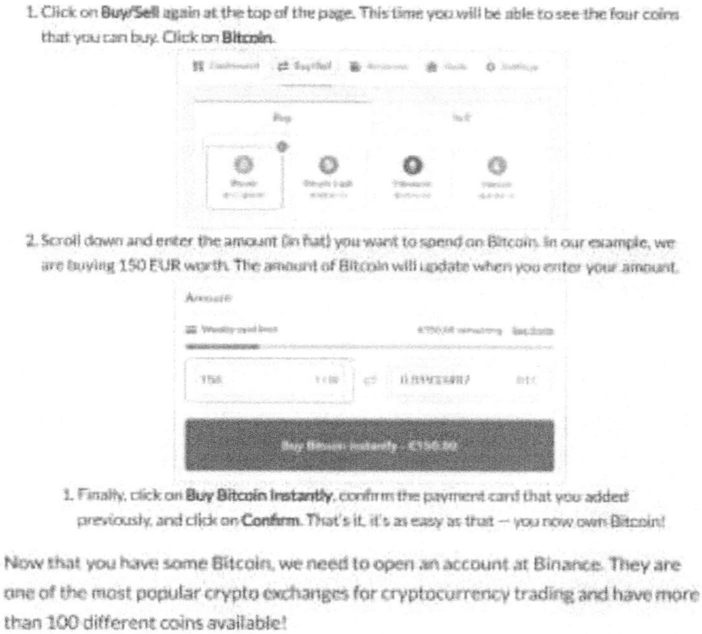

1. Click on **Buy/Sell** again at the top of the page. This time you will be able to see the four coins that you can buy. Click on **Bitcoin**.

2. Scroll down and enter the amount (in fiat) you want to spend on Bitcoin. In our example, we are buying 150 EUR worth. The amount of Bitcoin will update when you enter your amount.

1. Finally, click on **Buy Bitcoin Instantly**, confirm the payment card that you added previously, and click on **Confirm**. That's it, it's as easy as that – you now own Bitcoin!

Now that you have some Bitcoin, we need to open an account at Binance. They are one of the most popular crypto exchanges for cryptocurrency trading and have more than 100 different coins available!

If you want, you can obviously explore other platforms as well, but Coinbase seems to be one of the most beginner-friendly.

Step 2: Choose A Wallet

Cryptocurrency wallets are apps similar to those found on a smartphone or computer. If you prefer the tactile experience of holding a wallet, you can also buy a physical device that runs a wallet app

Satoshi Nakamoto created the first cryptocurrency wallet when he released the bitcoin protocol in 2009. Although Bitcoin is the most well-known and widely used cryptocurrency, others have emerged based on its blockchain technology. Any of them can be stored in a cryptocurrency wallet. Multiple cryptocurrencies can be stored in wallets.

When you want to acquire cryptocurrency, whether through a currency exchange or as a gift or revenue, you direct the sender to a unique cryptographic address issued by the wallet. You might imagine your cryptocurrency stored on the wallet in the same way that files are stored on a USB drive. Still, the information stored on the wallet only points to the location of your cash on the blockchain. This public ledger records and authenticates all cryptocurrency transactions. Spending with the wallet is as easy as scanning a retailer's QR code or directing a specific amount of cryptocurrency to the retailer's public address.

Some economists caution that cryptocurrency is a bad investment. You might be better off using Bankrate's investment rate comparison.

How Do They Work?

Crypto wallets, contrary to popular belief, do not actually store cryptocurrencies. They instead provide the tools needed to interact with a blockchain. In other words, these wallets can generate the information required to send and receive cryptocurrency through blockchain transactions. Such information includes, among other things, one or more pairs of public and private keys.

The wallet also contains an address, an alphanumeric identifier generated from the public and private keys. In essence, an address is a specific "location" on the blockchain to which coins can be sent. This means that while you can share your address with others to receive funds, you should never share your private key with anyone.

Regardless of which wallet you use, the private key grants access to your cryptocurrencies. As a result, even if your computer or smartphone is compromised, you can still access your funds on another device if you have the corresponding private key (or seed phrase). It is important to note that the coins never truly leave the blockchain; they are simply transferred from one address to another.

Difference Between Hot And Cold Wallet

As previously stated, cryptocurrency wallets can be classified as "hot" or "cold," depending on how they operate.

A hot wallet is any wallet that is connected to the Internet in some way. When you open an account on Binance and send funds to your wallets, you deposit into Binance's hot wallet. These wallets are simple to set up and provide quick access to funds, making them ideal for traders and other frequent users.

Cold wallets, on the other hand, have no Internet connection. Instead, they store the keys offline on a physical medium, making them resistant to online hacking attempts. As a result, cold wallets are a much safer way of "storing" your coins. This method, also known as cold storage, is ideal for long-term investors or "HODLers."

Binance only keeps a small percentage of coins in its hot wallets to protect users' funds. The rest is kept in cold storage and is not connected to the Internet. It is worth noting that Binance DEX offers an alternative for users who prefer not to keep their funds in a centralized exchange. It's a decentralized trading platform that gives you complete control over their private keys, as well as the ability to trade directly from their cold storage devices (hardware wallets).

The Different Types Of Wallets

Software Wallets

There are many different types of software wallets, each with its own set of characteristics. Most of them are linked to the Internet in some way (hot wallets). The descriptions that follow are some of the most common and important types: web, desktop, and mobile wallets.

Internet Wallets

Web wallets allow you to access blockchains via a browser interface without having to download or install anything. Internet wallets includes both exchange wallets and browser-based wallets.

In most cases, you can create a new wallet and secure it with a personal password. Some service providers, however, hold and manage your private keys on your behalf. While this may be more convenient for inexperienced users, it is a risky practice. You're entrusting your money to someone else if you don't keep your private keys. Many web wallets now allow you to manage your keys entirely or via shared control (via multi-signatures) to address this issue. As a result, it's critical to examine each wallet's technical approach before deciding on the best one for you.

When using cryptocurrency exchanges, you should think about using the available security tools. Device management, multi-factor authentication, anti-phishing code, and withdrawal address management are among the security features available on the Binance Exchange.

Desktop Wallets

As the name implies, a desktop wallet is a software that you download and run locally on your computer. As opposed to some web-based versions, desktop wallets give you complete control over your keys and funds. When you create a new desktop wallet, a file called "wallet.dat" is created on your computer and is saved locally. Because this file contains the private key information required to access your cryptocurrency addresses, it should be encrypted with a personal password.

If you encrypt your desktop wallet, you will be prompted for your password each time you run the software for it to read the wallet.dat file. You will most likely lose access to your funds if you lose this file or forget your password.

As a result, it's critical to back up your wallet.dat file and store it somewhere secure. You can also export the corresponding private key or seed phrase. You will be able to access your funds on other devices if your computer fails or becomes inoperable in some way.

In general, desktop wallets are considered safer than most web versions. Still, it is critical to ensure that your computer is virus and malware-free before installing and using a cryptocurrency wallet.

Mobile Wallets

Mobile wallets work similarly to desktop wallets but are designed specifically for smartphone applications. These are very convenient because they allow you to send and receive cryptocurrencies using QR codes.

As a result, mobile wallets are especially well-suited for daily transactions and payments, making them a viable option for spending Bitcoin, BNB, and other cryptocurrencies in the real world. Trust Wallet is a well-known example of a mobile cryptocurrency wallet.

Mobile devices, like computers, are vulnerable to malicious apps and malware infection. As a result, it is recommended that you password-protect your mobile wallet and backup your private keys (or seed phrase) if your smartphone is lost or damaged.

Hardware Wallets

Hardware wallets are physical electronic devices that generate public and private keys using a random number generator (RNG). The keys are then saved on the device, which is not connected to the Internet. Hardware storage is a type of cold wallet and is regarded as one of the most secure options.

While these wallets provide greater protection against online attacks, they may pose risks if the firmware is not properly implemented. Furthermore, compared to hot wallets, hardware wallets are less user-friendly, and funds are more difficult to access.

To circumvent this limitation, you can use Binance DEX to connect your device directly to the trading platform. Because the private keys never leave your device, this is a secure method of accessing your funds. Some web wallet service providers provide a comparable service, allowing hardware wallets to be connected to their browser interface.

If you intend to keep your cryptocurrency for an extended period or if you have a large amount of cryptocurrency, you should consider using a hardware wallet. Most hardware wallets currently allow you to set up a PIN code to protect your device, as well as a recovery phrase – which can be used if your wallet is lost.

Paper Wallets

A paper wallet is a paper on which a crypto address and its private key are physically printed as QR codes. These codes can then be scanned to carry out cryptocurrency transactions.

Some paper wallet websites allow you to download their code and use it to generate new addresses and keys while offline. As a result, these wallets are highly resistant to online hacking attacks and can be used in place of cold storage.

However, because of the numerous flaws, paper wallets are now considered dangerous and are discouraged. If you still want to use it, you must be aware of the risks. Paper wallets have a major flaw: they are only suitable for sending the entire balance at once.

For example, suppose you had created a paper wallet and sent multiple transactions totaling 10 BTC to fund it andIf you want to spend 2 BTC, you should first send all ten coins to another type of wallet (for example, a desktop wallet) and spend a portion of the funds (2 BTC). You can return the 8 BTC to a new paper wallet later, though a hardware or software wallet would be a better option.

Technically, suppose you import your paper wallet private key into a desktop wallet and spend only a portion of the funds. In that case, the remaining coins will be sent to a "change address" which the Bitcoin protocol generates automatically. You will most likely lose your funds if you do not manually set the change address to one that you control.

Today, most software wallets will handle the change for you, sending the remaining coins to a wallet-specific address. But keep in mind that your paper wallet will be empty after sending its first transaction – regardless of the amount. As a result, don't expect to reuse it in the future.

Some of the best wallets to check out are:

- Exodus
- Electrum
- Mycelium
- Ledger Nano X
- Trezor Model T
- Ledger Nano S

Step 3: Find Information Sources

Cryptocurrencies are highly volatile assets, with prices fluctuating by large percentages in a matter of hours. Because such movements can be triggered by various events, a cryptocurrency trader must constantly monitor market data.

Before investing in a coin, it is a good idea to research the asset. Coinmarketcap, a website where you can find information on every coin and token available, is one of the best tools for this research. The information includes capitalization, supply, trade volume, and so on. As a result, the site can be used to compare cryptocurrencies. Another useful site for this type of information is Cryptocompare.

The best way to stay up to date on crypto-related news is to subscribe to major industry publications and blogs, such as Coindesk, News Bitcoin, The Merkle, and Hackernoon. More useful information can be found on social media, such as crypto-related subreddits, Telegram and Discord channels, and Quora threads.

A CoinMarketCal, which tracks crypto-related events that may affect coins traded on exchanges, is another excellent data source.

Step 4: Use Technology To Enhance Your Chances Of Success

You should now be armed with some trading knowledge and tools. However, when compared to the fiat investment process, some elements are still missing.

To begin with, fiat investors can use stock indices as benchmarks when constructing their portfolios — the S& P 500 and Nasdaq Composite are perfect examples of such benchmarks. They enable investors to trade entire sectors and manage complex portfolios in a simple investment, lowering the portfolio's risks and volatility.

You should look for similar tools for the crypto market, such as Cryptoindex 100 (CIX100), an automated index calculated by a machine learning algorithm that analyzes cryptocurrencies. This tool enables traders to reduce portfolio volatility,risk,cost and effort of investing in multiple coins, resulting in significant trading savings.

Following the creation of the portfolio, you can track coins using specialized apps such as Blockfolio or Delta, which provide valuable insights (real-time exchange prices, the order book for every coin, etc.). Using such tools boosts productivity.

Chapter 7

Different Ways To Make Money With Cryptocurrency

When it comes to the best ways to make money online, cryptocurrency is probably not the first thing that comes to mind. However, it should.

Cryptocurrency is one of the most rapidly growing digital assets. It effectively offers multiple means for anyone looking to make a good income online, backed by the blockchain, and allows for trading.

There are numerous ways to make money with cryptocurrencies, some of which do not require you to invest anything other than your time and efforts.

Smartly Investing In Bitcoin

One of the best things about getting started investing in bitcoins is that a lack of previous investment experience doesn't hinder you nearly as much as it might in other markets, simply because most investors market are still so new. Additionally, due to the high level of volatility involved in the market, it is one of the most reliable investment strategies to pursue and the buy and hold strategy is one of the simplest.

To get started with a buy and hold strategy, all you need to do is wait until the price drops to its nearest current low and then use it as your entry point to hold onto the bitcoins until they reach a price near to their current high. With the profits you make, you then wait for the price to drop again and buy back in, repeating the process as needed to ensure that the number of bitcoins you accumulate will then likely be able to accrue enough value in the long-term to meet your investment goals.

Taking these steps will allow you to take full advantage of the concept of compounding, which is crucial to maximizing your investments in the long run while also minimizes the downsides that come along with investing in bitcoin, primarily the extreme amount of risk. The basic idea behind compounding is that reinvesting your early returns is the best and most effective means of maximizing your profits in the long run. Selling when the price is high and then buying again when it has dropped allows you to do just that.

To fully understand the potential that compounding offers, all you need to do is consider the example of the average 25-year-old who is looking to retire as a millionaire by the time he is 65. To do this successfully, he need to save an average of $900 per month, each month, until he retires, assuming he is seeing a very modest five percent return, per year, on his investment. However, if he waits just for 10 years until he is 35 to start saving, he would need to save about $2,200 each month to reach his goal. Likewise, if he waits to start saving seriously until he is 45, he would need to save around $4,500 each month to see the same results.

Long-term investment plans generally make do with reliability. Still, with minimal year-over-year returns, bitcoin's volatility means that depending on when you invest, you could easily see a decade’s worth of growth in a year, at least in the relatively near future. While these extreme pricing jags will naturally decrease over time as the price the market is willing to pay, and the price speculators have driven the price to come more in line with one another. As such, when it comes to investing in Bitcoin, there may very well never be a better time to do so.

While getting started quickly is key to maximizing your profits, this doesn't mean that you should jump in headfirst without stopping to think about what your personal investment habits are like to ensure that they end up helping, rather than hurting, your overall investment potential. When you are first starting out, the first thing you should keep in mind is that not all investing strategies will be right for every individual. The biggest determining factor is always going to be the level of risk that you will be comfortable with. While cryptocurrency is naturally riskier than a vast majority of other investments, this doesn't mean that there aren't still some shades of gray to consider on the topic.

Determining your true level of ideal risk should start with an honest assessment of the goals you have for the money you ultimately invest. Suppose you are more risk-averse, as a rule. In that case, you may be happy with the goal of keeping as much of your initial investment capital intact as possible. On the other hand, if you aren't afraid of pushing your risk to the limit, then your goal could be maximum profit regardless of the risk. Remember, when it comes to investing successfully, there is no risk without reward. If a greater risk doesn't come with a comparably greater reward, then it is likely not worth considering.

The specifics of the plan you choose aren't going to be nearly as important as the fact that after you decide as to how you are going to move forward, you stick to it no matter what. When an unexpected loss comes along to snatch defeat from the jaws of success, it can be easy to want to go against your plan and do something rash to regain lost profits; this is a negative habit to get into, and it will always lead to additional losses as opposed to profits. When you are putting your plan together, it is also important to keep in mind that your goals will not be completed in a vacuum; You will need to overcome many obstacles, and only hope to succeed once you have passed all the hurdles of life.

When creating a plan, the entire point is to use to make yourself as mechanical and logical in your investments as possible. Emotion is the enemy of logic. Giving into it will always cause you to see fewer profits than you otherwise would. A big part of keeping your emotions in check is ensuring that you never invest more money than you would be able to realistically afford to lose. Suppose you spend your time worrying that you will lose the money that you need to survive. In that case, there is no way you will be able to avoid the emotional pull to keep those funds safe, even if it runs counter to what your plan says to do.

When it comes to considering the right amount of risk to face regularly, as a general rule, the longer you have to plan on using your investment funds, the greater the amount of risk you can successfully take on at the moment as you have enough time to bounce back from any serious losses that come about as a result. All told, if you

are planning to invest for 10 years or more, then there is no reason you shouldn't lean as hard on potential risks as you possibly can.

Tips for investment success

Investing in bitcoins successfully is a skill that is only likely to improve with practice like any other. To speed things up, consider the following tips:

Always treat it like a commodity: While there is so much speculative interest around bitcoin these days, it is easy to forget that it is actually just a commodity, which means it is subject to many of the same rules as any other commodity. Like commodities such as precious or base metals, bitcoins are used for both speculative and practical purposes. As such, if you are interested in tracking the likely movement of bit-coin in either the short or long-term, the easiest way to do so is to look for ways that bitcoin is increasing its practical value as opposed to its speculative value. This means examining the aspects of the service that people are talking about and ensuring that public opinion (non-speculative, of course) remains strong. In the long run, what speculators think doesn't matter nearly as much as what the market believes the price is going to be worth.

Keep an eye on saturation levels: While Bitcoin is currently operating under a sustained pricing bubble, it is unlikely to burst seriously anytime soon. As such, there is still plenty of time to get in and make a profit at virtually any pricing point, as long as you are in a strong overall position when the mass saturation level reaches a point where it could tip over at any moment. When this occurs, it will be important to keep a close eye on what analysts believe the true market value of the cryptocurrency is. Suppose your investments were made above this level. In that case, you may be better off selling early instead of holding out to the last moment to maximize profits that may vanish if you aren't careful.

This mass saturation point is anticipated to take place sometime in 2022. However, the date could move forward if unanticipated leaps are made either in the technology itself or its adoption for one reason or another. This point will become closer and closer as new and improved ways of interacting with cryptocurrency are introduced to the public. They learn about the many ways in which blockchain technology can influence their lives for the better.

Know where you are in the market cycle: All markets and all assets move through the same cyclical pattern when given a long enough timeframe, which is what makes it so useful when it comes to pinpointing the current state of the cycle. The market cycle typically becomes noticeable when an asset reaches a state of stable growth that causes the market to feel optimistic about its potential. This naturally causes the price to rise even higher, which causes the price to rise to points that are ultimately unsustainable, thanks to euphoric investor speculation.

The state of euphoria will likely continue until the first shades of anxiety start to show as the bubble bursts and the price drops. This drop is then typically followed by a brief reprieve because investors refuse to believe what is happening right in front of them. This denial ultimately costs them, however, as the downward spiral is only starting. The price will continue to drop as the market moves from a period of fear to one of depression regarding the asset's prospects. When the point of no return is reached, panic is sure to set in, and the price will continue to decrease until it reaches its lowest.

With the price now again completely in line with what the market believes it to be worth, it will make it possible to begin to rise once again. This will, in turn, cause investors' confidence to rise from panic, back through depression, and eventually hope for the future.

Once hope sets in, it is only a matter of time before the market takes notice, and things start to become optimistic again .

Bitcoin has already been through the entire process once, finishing up with the crash of 2014. It is too soon to say, ifthe state of growth at the end of 2017 marked the start of the next true phase of optimism, or if it marks the start of the euphoric phase.. It is definitely something for investors who are just getting in now to consider, however, and worth keeping an eye out for the mass saturation point as well.

While Bitcoin is likely strong enough to survive any type of crash at this point, the next crash in price is likely going to take down a vast majority of the cryptocurrencies out there, more than 1,000 in all, in a purge akin to the dotcom bubble burst in the late 90s. Experts estimate that as little as 20 percent of the cryptocurrencies currently in operation will survive the mass saturation transition.

Plan for the long-term: While the growth Bitcoin experienced in 2017 turned many people into millionaires virtually overnight, it is important that you do not start investing in bitcoin to quit your day job within a year. While the 2017 price explosion wasn't the first time the price of a bitcoin has jumped substantially, virtually overnight, this in no way makes it anything but the exception to the rule. For each of the major pricing jumps, the price has stayed somewhat stable for years at a time.

This, in turn, means that you will need to approach investing in bitcoin with the right mindset to see sufficient results. Starting out with completely unrealistic expectations is only going to make it more difficult for you to stick to your plan and keep your emotions out of the equation entirely. Failing to do so can lead to a wide variety of potential issues that will make it difficult, if not impossible, to reach your long-term goals. While a vast majority of long-term investment options have additional lock-in risk associated with holding on to them in the long-term, this is one facet of investing in bitcoins that is less risky than the alternative. This is because unloading your bitcoins is as easy as going to any exchange and converting them back into cash. What's more, you can choose virtually any exchange that has the highest current rate as if you already have bitcoins in hand; most exchanges don't even require that you verify your identity before you are allowed to start trading. All told, this makes it easy to think of your cold bitcoin wallet as a type of savings account with an extremely high return on investment.

Cloud mining

Cloud mining is another type of long-term bitcoin investment option. In this case, you would pay a cloud mining service to mine bitcoins for you and then keep the profits. Suppose you don't have a significant amount of capital saved for investment purposes. In that case, this is a viable alternative as each transaction that is mined generates a small amount of bitcoin for the miner. Cloud mining works by sharing processing power to run data centers remotely that takes care of the actual mining. In this scenario, you don't need anything special other than a bitcoin wallet and a way to contact the cloud mining company you are considering.

However, cloud mining does come with additional risks, starting with the fact that many cloud mining services are currently scams, operating a variation of a traditional Ponzi Scheme. The best place to go when finding a legitimate cloud mining service will be the Bitcoin subreddit. Once you know the mining service is on the level, you will also want to be careful to read the contract closely, as some cloud mining services will require a certain commitment upfront. Others will shut down if the price of bitcoin drops below a certain point. Knowing what you are getting into is the best way to ensure that the process will be profitable for you in the long term.

All told, there are three main types of cloud mining services: hosted mining leases out entire mining machines and lets users to collect the profits. The next, virtual hosted mining creates virtual mining machines that run a portion of a machine's full power; inside these virtual machines, the user has complete control to mine in whatever way they see fit. The cheapest starter option is typically to lease a portion of a machine's hashing power and then generate profits based on what hashing power provides.

Mining

No, we're not talking about going into a cave with a pickaxe and a hardhat—we're talking about Bitcoin mining, which was one of the first methods of obtaining BTC. This is the process by which miners solve complex mathematical problems using powerful computers. They are rewarded with newly minted BTC when they crack the code. It's essentially a race to see who can solve the block fastest, and those who are fortunate enough to do so reap the benefits.

Mining wasn't always this complicated. Using only their home computers, some of the first miners were able to mine thousands of BTC. It's a completely different story now. To mine in today's BTC scene, you'll need top-tier equipment, which can cost several thousand dollars. Furthermore, because Bitcoin has made a name for itself, you will face a lot of competition.

Join Bitcoin mining pools or mining clouds to give yourself a fighting chance. A Bitcoin mining pool comprises miners who work together to pool their computing power to solve complex mathematical problems faster. A Bitcoin mining cloud is similar, but instead of connecting their computing power, they use the cloud to connect their computing power. They don’t have to install and run the hardware and software themselves. In either case, the miners' fees from transactions and the newly minted coin are divided among the group members.

you'll need better equipment to profit as the competition heats up. Furthermore, as you spend more money on equipment, your profit margin shrinks. As a result, Bitcoin mining is no longer as profitable as it once was. Before you start mining, do a lot of research because there are many easier ways to make money with Bitcoin.

Staking

Staking is the process of investing or locking up funds in a cryptocurrency to earn new cryptocurrencies in the form of interest. Furthermore, if you choose to hold your coins for a specific period, you will benefit from price appreciation.

Numerous proof-of-stake cryptocurrencies will allow you to earn a good living by staking. WISE, PIVX, Tezos, DASH, NAV Coin, and other cryptocurrencies are among the best.

WISE, for example, is an ERC-20 token that allows you to earn a handsome interest income by locking up your funds in the WISE contract. In addition to regular bonus offers, WISE allows you to earn up to 5% yearly interest on your stakes. WISE staking is safe, low-risk, and highly flexible, as you can withdraw or liquidate your stake at any time.

The Process Of HODLing

HODLing is the process of purchasing powerful cryptocurrencies with great concepts and good use cases and then holding them for an extended period in the hope that their prices will rise significantly. When the coin price is high enough for you to profit, you can sell it on the market. HODLing, on the other hand, does not earn you any interest in your money.

Cryptocurrencies such as Bitcoin, ETH, BNB, and others are ideal for long-term investment and HODLing.

Create Your Own Cryptocurrency

Finally, you can profit from cryptocurrencies by developing your own. Yes, you can create your own cryptocurrency, and it's not as difficult as it sounds.

Creating and launching your cryptocurrency, on the other hand, will necessitate some level of blockchain expertise. You must have a fundamental understanding of how blockchain and cryptocurrencies work. Then you must evaluate various options and decide on the type of coin you want to create. You could also hire a blockchain expert or team for the job.

Once your coin has been created, you can begin selling it through ICOs or exchanges. Of course, you'll need to promote your coin heavily at first so that users understand what it is and why they should invest in it.

Other Interesting Methods

Trading Your Crypto

Crypto trading works similarly to stock trading in which you buy and sell tokens/coins for profit margins. There are numerous exchanges where you can purchase the cryptocurrency of your choice. Then you can either hold them for a while or sell them as soon as the price rises.

There's also day trading for those who want to make a lot of money by selling and buying cryptocurrencies on the same day. You purchase a crypto-coin, wait for the price to rise, and then sell for a profit.

You can gain access to the project/product or use the coin for payments depending on the type of cryptocurrency you invest in.

Bonus Coins

Participating in bonus/airdrop offers is another good way to earn money with cryptocurrencies without any investment. All you must do to earn some free/bonus coins is join a cryptocurrency's airdrop or bounty program and complete some simple tasks.

You can also earn bonus tokens by investing in a cryptocurrency during the ICO or token sale phase. New projects frequently give bonus tokens to early investors.

Utilizing Referrals

A crypto referral program is a popular and effective way to earn cryptocurrencies. Most of the time, you won't even need to invest in anything to refer and earn money with a project. Simply register on the website to obtain your unique referral link and begin referring and earning.

WISE, like many other cryptocurrencies, pays well for referrals. WISE rewards both the referrer and the referee with 10% bonus tokens for each successful staking through referral.

Sell For Crypto

If you are a merchant and cryptocurrencies such as Bitcoin and Ripple are legal in your jurisdiction, you can accept them in exchange for your goods and services. Cryptocurrencies are digital money that can be easily transacted and are used globally without the need for currency conversion. As a result, for simple, global transactions, you can accept any of the top coins.

Several Bitcoin payment processors allow you to integrate cryptocurrency payment options into your offline/online store.

Dividends

Cryptocurrency dividends function similar to stock dividends. You invest in a dividend cryptocurrency and earn a fixed rate of return. You must purchase and hold cryptocurrencies for the duration of the interest-bearing period.

BNB, NEXO, and other top dividend cryptocurrencies which can be invested for interest are listed below. Most of these coins do not require you to stake to earn interest.

Earn By Working

As the popularity and reach of cryptocurrencies grow, an increasing number of businesses are now using them to pay their employees' salaries, among other things. Crypto payments are safe, inexpensive, and transparent. Furthermore, rather than sending cash through other methods, it is simple and inexpensive to pay your remote/freelance workers in bitcoin.

Whether you're a developer, a writer, or a designer, many international companies are willing to pay in cryptocurrencies for your work/services.

Chapter 8

Cryptocurrency Is The Latest Trend That Is Here To Stay

A Look Into The Future

Deutsche Bank is one of the latest investment institutions to throw its weight behind cryptocurrencies and Bitcoin. According to a report released by the bank, the current monetary system is fragile, thanks to the centralized nature of money and the growing unreliability of the system. Unreliability in this sense refers to the inability of money to provide basic security to most people.

The bank thinks that by 2030, digital currencies will have over 200 million users and these coins might eventually replace cash entirely. Demand for anonymity is growing, and governments worldwide have begun enacting privacy-centered laws that protect people from data exploitation. Cryptocurrencies satisfy this need, and there's no way paper money can ever do this.

Read an analysis of Bitcoin or cryptocurrencies. You'll often read predictions of how prices will hit $100,000 or some other astronomical number. These reports don't really highlight why or how a crypto coin will reach such a valuation. They instead focus on how much money people can make investing in them.

This isn't a bad thing by itself. However, it does set an investor up for trouble down the road. What if the coin doesn't reach its promised valuation? The investor will most likely panic and begin selling at the wrong time. Instead, imagine if they knew what was going on and could keep the bigger picture in mind.

Therefore, reading Deutsche Bank's report on this subject is illuminating. The bank draws interesting parallels between internet adoption and crypto adoption. Currently, cryptocurrencies are merely additions to the world's monetary system. By 2030, the bank thinks they'll replace existing monetary structures entirely as the users are going to nearly quadruple over the next decade.

This degree of growth is similar to how the internet grew from its beginning to what it is today. For those who remember, the internet was primarily used in an academic setting and was publicly funded. None of the original users ever thought that the network could be used beyond a single organization and connect people worldwide like it does today.

As policymakers and regulators freed it from private shackles, commercial activity increased. At first, no one knew what to do with the internet or even how they could connect to it. Soon, service providers began popping up everywhere, and before long, everyone could connect to it easily. As infrastructure grew, commercial activity and familiarity also grew, which fueled the rise and mainstream adoption of the internet.

Rising Tides Of Crypto

Examining the conditions that led to the rise of mainstream internet adoption and cryptocurrencies is instructive. There was massive value in creating policies that would have ensured full internet adoption. From increased connectivity to greater security, the benefits were evident. This was before anyone could envision how big the internet would become.

The cryptocurrency space is in the same moment right now. The failure of mainstream financial solutions is becoming more evident as each day goes by. Wealth inequality has been increasing in the Western world for a long time now, and routine fiscal policies have had no effect.

What's needed now is a catalyst or an a-factor' that will push cryptocurrencies from the fringes to mainstream use. Legislation is the only way forward, and the signs are good. The internet was legislated for mainstream use once companies began growing their offerings to accommodate new technology.

Much like how mainstream financial companies such as Paypal have begun offering crypto as a means of payment to their clients before legislation has occurred, companies began selling products on the internet before legislation arrived. If history is any indication, cryptocurrency regulation is just around the corner.

One of the things that push governments to legislate a new phenomenon is the rise of fraud. As more people jump into a new scheme or phenomenon, fraud rises. This is because actors of all kinds are attracted to the lack of rules and regulations . Cryptocurrencies have faced this issue for a while now. The early days of Bitcoin were marred by criminals using it to hide their transactions.

Despite the Federal Trade Commission busting many of these operations, the air or criminality surrounding Bitcoin has never fully dissipated. The truth is that Bitcoin is wholly unsuited to criminal activity because transactions can be traced easily. However, other forms of fraud have risen. From pump and dump schemes to fake ICOs, to multilevel marketing frauds that use the facade of cryptocurrencies, there's no end to how con artists design their schemes (Town, 2021).

As fraud rises, authorities will have no choice but to regulate the cryptocurrency space. At some point, rising fraud begins to harm larger portions of society, and this is where regulation must step in and "stem the rot,". In a thorough study conducted by the Deutsche bank, they were able to find that legislation is the only thing hasn't yet occurred in the field of cryptocurrency.

Everything else, from infrastructure to occupying mind space, has already occurred. It's almost as if we're waiting for the penny to drop. A pertinent question is: What happens once mainstream legislation is passed? Will prices drop or rise? There's no easy way to answer this question.

A lot depends on the coins themselves and the perceptions surrounding them. There's no doubt that regulation will open the doors of investment into coins like Bitcoin and Ether, which will send them skyrocketing. However, will this rise sustain itself for over two to three decades? No one knows.

Fake currencies and altcoins that no one has ever heard of might be doomed. Holders of coins such as Doge might find themselves with lot of money. There's no doubt that people pushing pump and dump schemes on social media will face harsh penalties due to their actions. The dawn of regulation brings more attention to an asset's fundamentals, and this will weed out the less-than-spectacular coins.

The biggest beneficiary of regulation might be Ether. Ether's blockchain has practical use and is the most widely used for smart contracts and decentralized finance applications. Ether as a token has true value, and regulation which will remove the cover on the fundamentals of this coin.

Bitcoin could go either way. On the one hand, greater mainstream adoption will push its prices higher. Everyone has heard of Bitcoin, and if the authorities announce regulation and oversight, people will have greater confidence to jump in, which will push prices up. However, increased scrutiny on the fundamentals of Bitcoin might reveal that there are better coins out there, or coins that are more in tune with investor goals which are deflating prices.

From a fundamental investor's perspective, it's hard right now to calculate what the value of a coin ought to be. Should they be treated as currencies, which are essentially commodities and don't have intrinsic value, or assets with real value? Cryptocurrencies are an exception because they lie at the intersection of an asset and a commodity. They are coins that are used in transactions but also have asset-like scarcity qualities. For instance, real estate is automatically valuable because it's scarce, and everyone needs it.

Similarly, Ether is valuable because people need smart contracts, and the coin is scarce. Bitcoin is scarce and also has value because it provides people with an alternative means of wealth storage.

Sustainability

When looking at Altcoins and investment opportunities in general, the underlying sustainability will have an ever-increasing impact on the market.

In March 2021, Elon Musk announced that the electric carmaker Tesla would allow customers to complete Bitcoin purchases. A few weeks later, in May 2021, this initiative was reversed, with Elon citing the environmental concerns associated with Bitcoin. These two announcements contributed first to double-digit percentage gains in the crypto market, followed closely by even greater declines, as much as 30% in a day. The fall was compounded by a Chinese government

crackdown on banks' use of cryptocurrencies. Should this be a time to panic or an opportunity to double down on your investments?

In March 2021, the annual power consumption of the Bitcoin network was estimated to be 129 terawatt-hours (TWh). By contrast, this is greater than Norway (at 124 TWh, with a population of circa five and a half million) and two and a half times the energy use per year of Amazon, Apple, Facebook, Microsoft, and Google combined.

As of May 2021, Ethereum has stated it is just "months" away from migrating from a proof of work to a proof of stake protocol. This, they claim, will slash its carbon emissions by hundredfold. Such a migration would lead to increased market confidence in Ether. At the same time, proclaimed sustainable currencies, such as Stellar and Cardano, would similarly be well-positioned as environmental concerns and will become pivotal in investment strategies.

A Solution

The biggest argument against crypto going mainstream is that governments aren't going to give up control of fiat money without a fight. This is true. For instance, it's hard to imagine the United States government giving up control of the USD and allowing people to trade Bitcoin without any control over the supply of Bitcoin.

While this attitude is often presented as a bad thing, it's good that governments control currencies. This keeps the economy stable and prevents a single malicious actor from cornering a coin and manipulating the economy. There are many conspiracy theories out there of powerful people who control the world's economies or the Illuminati or whatnot.

A currency that isn't under government control will make such a situation a real possibility. Thus, it's in the best interests of society that finance doesn't become fully decentralized. However, the current emphasis on centralization is also not working. A middle ground between decentralized and centralized finance is the way to go, by all accounts.

Many commentators have speculated that fiat money will come to an end. This is highly unlikely because of the reasons I've just outlined. However, the end of fiat currencies as the primary means of transaction isn't implausible. In fact, it might be the best compromise in this situation.

Stablecoins, or coins that tie their value to fiat currencies, are a possible solution. These coins give people all the benefits of a cryptocurrency

. Still, their values can be controlled by the issue and withdrawal of fiat currency. Thus, governments can exercise control. People can claim the control they've ceded in the past to the centralized financial model.

The real victim of all this will be plastic cards. In fact, digital modes of payment have eliminated these cards for the most part. In China, the rise of WeChat (a super app with all the facilities one might need) has eliminated the need for a card. Similarly, Singapore-based Grab is bringing the same degree of convenience to users in Southeast Asia. India's PayTM, a mobile wallet, is rendering both cash and cards obsolete. It also allows their people some solace amid its questionable government policies.

Where does crypto fit into all this? Well, for starters, integrating these solutions with a digital currency is easy. Enforcing KYC norms can be done via smart contacts.

These mobile payment options will only increase security and the confidence that people feel.

When the internet was first adopted, we used it to primarily browse websites. No one could have imagined that someone could walk into a cashier-less store, leave without paying, and handle payment via an application on their phone. Who would have thought that we could simply wave our phones over a card machine and pay for purchases? Or that wallet sizes would shrink immensely?

Similarly, we don't know how the crypto field will play out. Who knows what apps will emerge once crypto goes mainstream? As an investor, the best position to adopt is to HODL for as long as possible. The inherent value in these coins will make itself known and will eventually bring profits to long-term investors.

Thus, there's no need for anyone to panic or worry about which way the market will go in the short term. Returning to fundamentals is the key. Hopefully, this discourse has helped you figure out what the deal is with cryptocurrencies.

The Most Promising Coins To Buy Now

Cryptocurrencies have seen their biggest surge since late 2017, attracting a lot of attention. Investors are looking for the top cryptocurrencies to diversify their portfolios. Bitcoin, the most popular cryptocurrency, recently hit a record high of $63,000 before experiencing high sales, providing investors with an opportunity to purchase bitcoins before the price spikes again. One bitcoin is now worth $56,000.

In recent months, cryptocurrency prices have fallen from their highs after US President Joe Biden signaled plans to raise capital gains taxes, prompting some investors to sell their holdings. However, this hasn't stopped financial services from being bullish on digital currency.

Visa announced at the end of March that it will accept USDC Stablecoin for transaction purposes. The Gemini cryptocurrency exchange announced April 2021 that it will partner with Mastercard to launch a rewards-based credit card. Users will be able to earn cashback in the form of cryptocurrencies. JP Morgan is also planning to launch a managed bitcoin fund for its clients. Not only bitcoins, but the meme cryptocurrency dogecoin, which has been making the rounds on social media, is now being accepted by merchants such as the online tech retailer NewEgg, The Kessler Collection luxury hotel chain, and the Dallas Mavericks NBA team. Not to mention, Elon Musk's Tesla

Inc. will begin accepting bitcoin as payment for electric vehicles.

Ether

Ether is the second-largest cryptocurrency created in 2015, with a market capitalization of $316 billion. The Ethereum blockchain is based on smart contracts, which enable decentralized applications such as the well-known decentralized finance services (DeFi). While investors view bitcoins as a form of gold, Ether's feature allows for a greater number of public blockchain transactions.

The current value of Ether in USD is $3,376.34.

Binance Coin

This is the third-largest cryptocurrency launched in 2017, with a market capitalization of $90.9 billion. The coin is exclusive to the Binance exchange and the company's new ecosystem, Binance Smart Chain. The company, which is one of the largest in the world, aspires to become the primary infrastructure service provider for the entire blockchain ecosystem.

The current USD value of the Binance coin is $663.43.

Solana

Solana is one of the fastest-growing DeFi ecosystems. It now has its own cryptocurrency coin, which peaked last month. Launched in March 2020, it has quickly risen to the 14th largest cryptocurrency, with a market capitalization of $11.19 billion. Its Proof-of-History algorithm has the fastest transaction speeds. In comparison to Ether, Solana has lower transaction fees.

Solana's coin is currently worth USD 47.07.

Dogecoin

This is the most affordable cryptocurrency to purchase in 2021. Because Elon Musk refers to himself as the "Dodgemaster," this coin has risen to prominence after a shady past. Nobody expected this coin, created in 2013 as a meme coin, to become legit, let alone be included in the top list of cryptocurrencies. In less than two weeks, the price jumped from $0.007635 to an all-time high of $0.074398. All of this happened in the aftermath of the GameStop incident and Elon Musk's tweets.

The current value of Dogecoin in USD is $0.4392.

Cardano

Cardano has promising growth potential. The coin's value has risen from $0.182 in January to $1.45 in February, making it the most cost-effective cryptocurrency to invest in. The recent update in the code of Cardaon is responsible for the increase in value, which added the ability to create smart contracts, allowing developers to begin working on dApps, many new cryptocurrencies, and the DeFi sector. This coin also has low transaction fees and a strong development team working to fully decentralize the project.

Cardano is currently worth $1.32 in USD.

Rising Trends In The World Of Cryptocurrency

Cryptocurrency trading has risen to prominence among investors and traders. The Crypto converter is a tool for determining the price of Bitcoin in USD. Investors and traders have the best opportunity to visit a trusted and reliable online store and proceed with step-by-step integration of plans to demonstrate your interests and trust levels on behalf of smart choices. Use the best-recommended crypto converter to learn about BTC to USD because only an authentic and reliable Bitcoin converter can help you achieve your goals and make the right steps at the right time for interested communities. Get instant access to a trusted crypto converter. Because of numerous risk factors, it has been determined that cryptocurrency should account for no more than 3% to 5% of total assets.

Bitcoin Convert & Automated Crypto Converter

It is not difficult to calculate the Bitcoin price; simply visit a live Cryptocurrency platform to view the most recent BTC to USD exchange rates. The 1 BTC to USD converter provides instant access, and smart feature plans to help you move forward with online resources that are easily accessible. Conversion rates from digital currencies to other currencies can be found on authentic and reputable websites such as https://ccprice.com/. Get immediate access to the Cryptocurrency Converter and Calculator Tool and ensure how to satisfy and match your preferences to get the best and authentic source. Crypto exchange rate calculator assists you in converting prices so that you can proceed to approach instant and dependable resources. The simplest way to obtain live exchange rates is to have genuine ideas and useful acknowledgment to proceed with dependable and quick accessibility resources. Determine which tool is the best and how to be satisfied to calculate the BTC Value.

The Most Profitable Online Opportunity Tool

Non-fungible tokens (NFTs) are Blockchain-based tokens with a high reputation and demand based on cutting-edge technology. No one can deny the significance of NFTs tokens in new Cryptocurrency trends. There is a vast array of ideas and useful strategies that can be combined and chosen to make money online. Crypto trading and parameters encourage investors to proceed with an instant and reliable source of action plans, proceed with reliable sources, and ensure smart feature choices in the latest trends 2021.

Trends in Cryptocurrency Investment in Covid-19

Cryptocurrency investment plans are regarded as one of the best and smartest feature investment plans. They are in high demand among traders worldwide. People want to see the most recent Cryptocurrency exchange rates with USD and compare their currency values to ensure their profits. Different modes of Cryptocurrency investment trends have different ratios that can be pursued and can be approached using the best and most trustworthy resources. When compared to other digital currencies, Bitcoin has the highest liquidity value.

Different Forms Of Digital Currencies

Some of the best digital currencies are TRON, Ripple, Cardano, Ethereum, Binance Coin, and Dogecoin. With Bitcoin's high liquidity, the trend of investing in online digital currency has grown in popularity. The trend of making online investments has increased as a result of the Covid-19 situation. March 2021 was the peak month for Bitcoin, with impressive investment levels from investors worldwide and many new investors investing in online digital currency.

Organized Selection Of Currencies

Before making any investments, use the Crypto exchange rate calculator to analyze the currencies and their current values. People can learn about the comparison of two currencies by converting prices online between two currencies. Proceed with immediate and dependable resources and ensure how to be satisfied and match your preferences to proceed with genuine online resources. Compared to other forms of currency, Bitcoin has a tremendous status and a high reputation for online money-making ideas in the long term. The selection of the best forms of digital currencies from the newest and fastest-growing Cryptocurrencies has great value and useful inspirations to move forward with the right time steps and increase the credibility of the investor's profits through genuine resources.

Integration of Plans in Steps

Bitcoin reaches record-high values due to investor confidence and the acquisition of the best and smart feature plans to proceed with step-by-step integration of plans. Bitcoin investors can only purchase within a certain price range. Get useful recognition and useful strategies to learn about useful facts about Bitcoin and its other parameters to proceed with the right time steps. Examine the live Bitcoin Currency process and then proceed with the step-by-step integration of plans to achieve your goals. Check out the most recent price comparison tool for the best opportunity to carefully compare prices online. Make sure how many levels of investment can be proceeded, but careful analysis is required before proceeding.

Guaranteed Converter Online Access

Determine which patterns and work plans are required and achieve your goals to find the most authentic and genuine resource. Make sure you understand how to use it because it is much simpler and easier to use than an online resource. There is no complication factor involved in using the Bitcoin converter by having a free official account. Numerous other online Bitcoin calculator tools can be used and accessed but having online fast accessibility features plays an important role in enhancing profit-making ideas and increasing real money value.

Why 2021 Is A Big Year For Cryptocurrency?

It is always exciting to forecast the trajectory of the crypto industry because it is developing both with conjunction and in opposition to the traditional financial system. On the one hand, the introduction of requirements for crypto user identification, the growing interest in digital government currencies, PayPal's crypto service, and the upcoming launch of the stablecoin, Diem (ex-Libra), from Facebook, among other events, confirm that digital assets are finally becoming more understandable and becoming mainstream.

Will Determine The Value Of A Lot Of Major Coins

The speed of cryptocurrency distribution, on the other hand, is directly related to how quickly operations with their various brands and flavors become available and accepted in each traditional bank or payment system. The widespread use of digital assets is both a goal and a source of concern for the world. The attempt to strike a balance between profit and risk in cryptocurrencies will determine 2021 trends.

Crypto Will Observe Some Form Of Tax Regulation

The main topic for the near future is cryptocurrency taxation. Today, crypto taxation is still a mystery – an idealized picture which is far from reality. Crypto taxes are not yet widespread. While they are unpopular with some, they have begun to appear in some countries as the markets mature. Governments see the potential revenue raisers outweighing previous crypto uncertainties.

However, implementing mandatory user identification through knowing your customer (KYC) procedures, the development of protocols that enable transaction tracking, and the adoption of legislation on digital assets indicate that things are changing and changing faster than some might expect.

We also see governments exchanging information on cryptocurrency owners and transactions and monitoring tools being actively developed. As a result, the world is likely to witness the first bitcoin tax evasion lawsuits in 2021.

Silent Crypto Harbours Will Make An Entrance

Because there is an anti-trend for every trend, the imposition of cryptocurrency taxes will increase the attractiveness of jurisdictions that will resist this practice and allow users to legally reduce the costs of owning digital assets. Simply put, "offshore crypto havens" will become more active. Countries with well-developed IT and financial markets, such as Singapore, Korea, Japan, and, of course, Switzerland, will most likely play this role.

Oncoming Crypto Crisis

The maturing cryptocurrency world is not only becoming more transparent, regulated, and secure, but it is also beginning to face a variety of economic challenges and tests. We are already witnessing the first signs of a crisis unrelated to cybercrime or fraud.

Bitcoin (BTC) reached a new high in December, surpassing the $34,000 mark. However, the reason was increased demand for BTC and an oversupply in the market of stablecoins Tether (USDT), which are used to conduct 70% of crypto exchange trading.

Tether, which is registered in the British Virgin Islands, is constantly increasing its coins' emission to increase the capitalization of its coins. At the same time, market participants are skeptical that USDT stablecoins are truly backed by fiat assets, such as US dollars. Furthermore, Tether is owned by the company iFinex, against which investors filed a $1.4 trillion class-action lawsuit in 2017-2018 on charges of market manipulation.

As a result, we see that crypto exchanges today is roughly like the traditional economy where governments start printing mony: an excess of fiat money supply in the market leads to dollar inflation and thus devaluation. We see the depreciation of money, which is currently USDT in the crypto world, leading to an increase in the cost of goods, which is BTC in the crypto world. As a result, current trends may result in further depreciation of altcoins and an increase in bitcoin price, whose supply is well known to be limited.

Risk Assessment Systems Will Improve

With the value of bitcoin increasing, there is an urgent need to develop a high-quality risk assessment model, as it is becoming increasingly difficult for users to objectively assess the potential outcome of crypto investments without succumbing to the general rush. Services that provide a working solution, rather than just "digital fortune-telling on the coffee grounds," will be able to quickly win the hearts, minds, and wallets of both new and experienced cryptocurrency market participants.

Today, there are over 8,000 different cryptocurrencies in the world, according to CoinMarketCap. More than 90% are fraudulent schemes, also known as "scams," in the industry. However, many of the remaining 10% have growth rates comparable to, if not better than, Bitcoin.

At the same time, those planning to invest in cryptocurrency must consider several risks that can cause the value of a coin to rise or fall:

The value of the currency might depend on the rules and regulations of the country/organization it is originating from. for example, which country the issuing company and the crypto exchange are based in, and what legislative changes are being implemented in that country in favor of or against digital assets; these factors might cause significant fluctuations in pricing.

Technical flaws in the code, insufficient information security, and insufficient data protection can all be exploited by cybercriminals to steal cryptocurrency.

Price risks continue to be the most difficult to assess. Analysts can track the movement of significant volumes of cryptocurrencies, determine who owns them, and observe actions related to their sale, thanks to the ubiquitous KYC (user identification) and KYT (transaction identification) rules. Based on the data obtained, predictions about changes in the value of the cryptocurrency can be made based on the goals, time, and other characteristics of such sales. The increased market size also reduces the market's reliance on individual speculation.

There is less uncertainty in the crypto world today, and more opportunities exist for developing analytical tools. However, novice investors continue to struggle with understanding the complexities of alternative finance.

Services that provide a working solution, rather than just "digital fortune-telling on tea leaves and coffee grounds," will be able to quickly win the hearts, minds, and wallets of both new and experienced cryptocurrency market participants.

Transactions Fees And Costs Will Alter

This trend is intriguing because it will have multiple directions. Because of technological advancements, ether transactions will become less expensive, while bitcoin transactions will continue to rise in price.

Changes in operating costs can have an impact on e-commerce players' interest in cryptocurrencies. Today, acquiring crypto attracts online stores because it is significantly less expensive than fiat currencies. The ability to maintain this advantage, in the long run, will largely determine the rate at which cryptocurrency spreads as a payment method.

Rise Of 5G Will Transform The Market

Many people still underestimate the 5G standard, which is a new paradigm in data transmission. Its implementation will result in the emergence of new concepts, types of services and changes in how mining is built, type of DeFi applications in development, and other factors.

Transaction management capabilities will no longer be limited by network data speeds with 5G. For example, when investment decisions are made by computers, 5G can significantly impact the high-frequency trading segment, especially with the ultra-low latency that 5G provides.

Traders are currently attempting to locate their server as close to the crypto exchange as possible because the length of the wire affects how quickly they can place or withdraw an order. 5G will help overcome this barrier by providing a level playing field for transactions across all systems, regardless of where the crypto exchange is located.

Things happening in front of our eyes is something that skeptics thought was impossible until recently: the world of finance has become multipolar. Regulators, traditional financial institutions, and crypto companies are increasingly collaborating to maximize the benefits of crypto technology. Although not all major issues have been resolved today, I am confident that many of them will be resolved by 2021. As crypto continues to mature and gain global acceptance, a positive outcome is unavoidable.

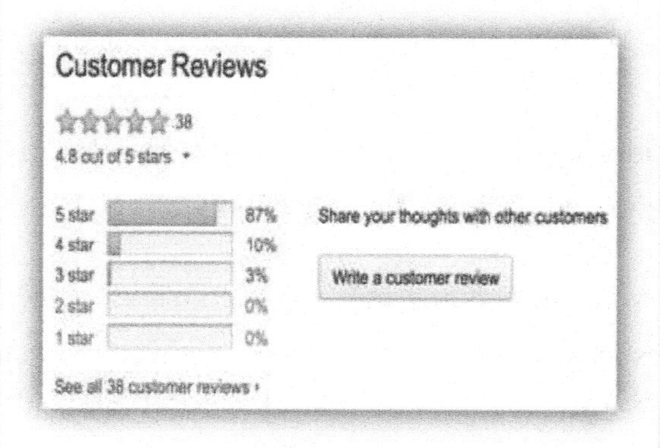

Please Leave a Review!

I would be incredibly thankful, if you could take just 60 seconds to write a brief review on Amazon. Even it's just a few Sentence!

Conclusion

Kudos to you if you have read this book through the end and grasped the concepts within.

We all want to become insanely rich, right? But the path to riches and grandeur isn't easy. It is paved with stones of hardships. But with sheer willpower and dedication, it is very much possible to reach that point where you can just sit back, relax and enjoy life as it was meant to be!

However, with the advent of the internet and the countless opportunities to make money, the process has become a little bit easier than before.

I certainly hope that the information within this draft was informative and provided you with all the tools you need to achieve your goals.. However, this is just the beginning! There is much more to learn on cryptocurrencies, so please continue to learn by looking at other resources, talking to fellow investors, and looking at reputable sources for advice.

Remember, for all the cultural cache that bitcoin has now, the fact is that it is still an extremely new technology, which means that dedicating yourself to the market study is the only way to see reliable results in the market long term. As such, after you start investing or trading, resting on your laurels is only going to see you losing out in the long-term to those who were better informed. Keep it up, and you will need a bitcoin money bin to keep all your bitcoins in.

Likewise, it is extremely important to take your time when creating your personalized trading plan. While it is also important to not wait too long to take full advantage of the potential for profit bitcoin offers, there is nothing to be gained by jumping in early without a firm idea of what you hope to accomplish. Taking a slow and thoughtful approach will allow you to maximize the advantages you do have, leading to better overall profits in the process. However, this is not to say that the going will be easy as the volatility in play means nothing is certain. However, you can count on that making money from bitcoin is a marathon, not a sprint, which means that slow and steady wins the race.

As a final request, I would encourage you to leave feedback for my book on Amazon if you find this book helpful and effective. It would mean a lot to me and inspire me to produce more high-quality content for you in the future!

Stay Safe and Stay Blessed.

Made in the USA
Monee, IL
08 November 2026

67378334R00121